Who Approved This?

Enterprise AI Governance and the Decisions Nobody Owns

BANDAR ALSHARFI

CODEDEBATE PRESS

Published by CodeDebate Press, an imprint of RB-Ventures LLC.

ISBN 979-8-9973353-0-4 (e-book)

ISBN 979-8-9973353-1-1 (paperback)

Quantity sales. Special discounts are available on bulk purchases by corporations, associations, and educational institutions. For details, contact the publisher at codedebate.com.

First edition
CodeDebate Press · codedebate.com

In memory of my mother.
She gave up more than she ever let us see, and asked for nothing back —
the first person I ever watched take responsibility for something
simply because it needed doing.

Everything here began with her.
I miss her, and I hope she would be proud.

For Rania,
who holds the real decision rights in this house
and has never once needed a framework to exercise them.

For Ahmad,
who reviewed none of this, approved all of it,
and remains my most enthusiastic stakeholder.

And for Mo, my partner in enterprise AI,
who has heard every argument in this book at least twice
and made both versions better.

Whatever else is unowned in these pages,
this one has four names on it.

Why This Book Exists

Enterprise AI does not fail because the technology doesn't work. In organization after organization, the models are accurate, the platforms are capable, the teams are good. The failure comes later, and it is quieter. It arrives the moment someone in a boardroom asks a question that sounds simple: *who approved this?* And then silence, because there is not one answer but too many, all contradicting each other.

That silence is the subject of this book. It is the sound of governance architecture breaking under the weight of something genuinely new.

Here is the core argument. Traditional governance works because it assumes clear ownership boundaries: one application owner, one security review, one compliance sign-off, a chain of custody where every link has a name. AI breaks that assumption. A single model's behavior depends on data quality, training methodology, business-logic interpretation, deployment infrastructure, vendor dependencies, and ongoing monitoring. Five or six domains that don't normally talk to each other, each able to point at a piece it doesn't own. So enterprises reach for the governance they already have, the governance built for traditional software: a review board, a checklist, a policy document, a committee. And they get governance theater: plenty of process, plenty of documentation, and no one actually accountable when it goes wrong.

The mistake is treating governance as a checklist when AI demands an operating system. A checklist is a static list of things to verify once, before launch. An operating system is a small set of boring, load-bearing decisions embedded in how work actually happens, and sustained after launch: who decides, what they are accepting when they decide it, what triggers a reversal, who watches the signals, and how an alarming observation reaches someone with the authority to act — fast enough to matter. The distinction is not academic. A checklist tells you the model was validated. An operating system tells you whether deploying it was a good business decision, who made that call, and what happens when reality shifts.

This reframe has an uncomfortable implication and a reassuring one, and they are the same one: the work is organizational, not technical. You cannot buy your way out of it. No vendor's platform will solve it. What actually works is specific: name a single business decision-maker who bears the downside, feed them real perspective instead of false consensus, document exactly what was accepted, and give that person standing authority to halt or redirect. It is the kind of work that never makes the conference keynote, which is precisely why the organizations that do it pull ahead.

This book is the field manual for that work. It is built from ninety essays written in and around real enterprise deployments, and it is organized the way the problem actually presents itself: from *who owns this?* through model risk, vendors, operations, economics, and regulation, to the point where the pieces cohere into one system you can run. It will not make governance exciting. It will make it real.

Who This Is For

This is a practitioner's book, written for the people who have to answer for AI when it matters — not the people who build the models, but the people who are accountable for using them.

The board member and the CxO. You are handed a model-accuracy dashboard built for data scientists and asked to exercise fiduciary oversight with it. You can feel that it's the wrong instrument (nobody can answer when someone asks what could go wrong), but you haven't been given the decision-grade information, the risk-appetite language, or the governance structure that would let you oversee this the way you oversee everything else. This book gives you those.

The Chief Risk Officer and the head of governance. You have been asked to govern AI without the budget, the technical fluency, or the position in the decision flow to actually prevent a bad deployment. You are an auditor when you need to be a governor. You will find here the operating model that moves governance upstream and puts real authority where the decision is made.

The General Counsel. Your worst exposure is not the dramatic failure that ends in a lawsuit; it is the quiet harm that, on inspection, nobody actually decided, plus a stack of training data whose rights you can't defend. This book reframes liability, IP, and compliance as one discipline: a named human owning a documented decision before deployment.

The head of AI, data, or the program itself. You are closest to the technology and furthest from the authority to own its consequences. You have watched governance become a bottleneck teams route around, or a fragmentation that spreads risk everywhere. You want speed and safety to stop being a trade-off. That is the entire project of this book.

The moment you are all in is the same moment: AI has stopped being a pilot and become infrastructure. It is making decisions that cost money, touch customers, and create exposure — and your governance was designed for a world where one person owned each decision. This book is for the leader who has noticed the gap and refuses to paper over it with another committee.

How to Use This Book

There are three ways to read this book, and they are not mutually exclusive. Most readers use all three over time.

1. Cover to cover — the argument, in order. The book is a single sustained case, built in eight parts that follow the problem as it actually unfolds: from *who owns this?* to model and data risk, to vendors, to operations, to economics, to regulation, and finally to the systems view where the pieces become one operating system. Read straight through, it is an argument for governance as organizational design. If you read nothing else, read Part I and Part VIII: the frame and the synthesis.

2. By pillar — reference when you hit the problem. Each chapter is self-contained and maps to one of the fifteen governance pillars. When your board asks for a risk-appetite statement, go to Chapter 5. When your budget blows up, go to Chapter 15. When a regulator sends a letter, go to Chapter 19. You do not need the preceding chapters to use the one you need. The annotated table of contents below is built to be scanned exactly this way.

3. As workshop material and enterprise artifacts — the part that earns its keep. This is the use we most want you to make, and the one most readers overlook. **Every chapter ends with a Field Kit**, a set of numbered, workshop-ready tools you can put in front of a team on Monday morning: the five questions to bring to any deployment decision, the ownership-vs-accountability audit, the decision-record template, the operational-readiness audit, the sunk-cost circuit breaker, and dozens more. These are not summaries. They are the artifacts. Pull the relevant Field Kit into your next architecture review, risk committee, or governance offsite, run the exercise as written, and let it surface the disagreement your consensus was hiding.

Every chapter's Field Kit works as printed; you can run any of them straight from the page, and nothing you need is held anywhere else. The visuals are built to the same standard: the decision-rights maps, the monitoring-layer diagrams, the maturity ladders and the fifteen-pillar schematic all reproduce legibly on a whiteboard or a slide. Use the prose to understand *why*; use the Field Kits and the visuals to make it happen in a room full of people who have to decide. A chapter you agree with changes your mind. A Field Kit you run changes your organization.

The Argument in One Page

AI becomes a governance problem the moment nobody can answer *who approved this?* — because it distributes responsibility across data, models, vendors, and decisions that your traditional governance assumed one person owned, and the fix is organizational, not technical. Someone therefore has to *actually decide*: place each decision at the level that can make it fast enough and own it, put one named person in the seat with reversal triggers, build accountability that learns instead of scapegoats, and give the board an oversight instrument it can act on. Accuracy tells you a model works, not whether deploying it is a good decision, and the real risk surface is the data beneath the model and behavior you can no longer hold still, so you govern decisions and provenance, not just metrics. Building your advantage on a vendor's model is a strategic dependency few boards have priced, and independence turns out to be an architectural property you design in one decision at a time, not a vendor you pick. In operation, a technically healthy model can make worse decisions for months, so you monitor behavior rather than uptime, design for the silent failure, and remember that your most successful model is your most concentrated risk. The economics work only when you control cost with unit economics instead of forecasting it, build capability instead of hiring a unicorn, and measure whether people actually decide differently rather than whether they log in. Regulators audit organizational judgment, not architecture, so the durable move is decision infrastructure plus modular compliance that flexes across contradictory jurisdictions rather than chasing today's rules. Finally, the fifteen pillars are one operating system that must be realigned at the seams and rebuilt at each scale threshold. And built deliberately, governance stops being friction and becomes the advantage that lets you move fast with confidence.

Table of Contents

PART I

Why AI Is a Governance Problem, Not a Technology Problem

CHAPTER 1

The Ownership Question

Part I · Why AI Is a Governance Problem, Not a Technology Problem

I watched it happen in three different organizations within a single year. Each one had done the hard work — hired the right talent, built capable systems, deployed models that worked. Then came the moment when someone in the boardroom asked a question that sounded simple: "Who approved this?"

The room went quiet. Not because they didn't have an answer. They had too many answers, all contradicting each other.

In one case, the answer was the VP of Analytics. In another, it was Product. In a third, it was "well, the data science team built it, and operations deployed it, so…" That trailing off matters. It's the sound of governance architecture breaking under the weight of something genuinely new.

This is the moment your AI strategy became a governance problem. And here is the part most enterprises get wrong: it did not happen when you deployed your first model. It happened when you assumed your existing governance structure would work for something fundamentally distributed across your organization. When "who approved this?" has no single answer, that's when you know you're there.

Here is the reframe this chapter is built on.

The myth: AI governance is a technology problem, and the governance structure you already have will work once the models are deployed.

The reality: it is an organizational design problem, because AI distributes one decision across five domains and leaves you plenty of ownership and almost no accountability.

The reassuring news, and the uncomfortable news, are the same news: this is not a technology problem. It's an organizational design problem. You cannot buy your way out of it, and no vendor's platform will solve it for you. But it is solvable.

Where Clear Ownership Breaks

Traditional governance works because it assumes clear ownership boundaries. The application owner is responsible. The security team reviews. Compliance signs off. Audit checks the box. There is a chain of custody, and at every link you can name the person holding it.

AI breaks this. Not because AI is magic, but because it distributes responsibility across domains that don't typically talk to each other. A model's behavior depends on data quality (traditionally data engineering's problem), training methodology (data science), business-logic interpretation (product), deployment infrastructure (operations), and ongoing performance monitoring (analytics, or sometimes risk management). Five domains, five owners, one outcome.

So when something goes wrong (a model drifts, produces unfair outcomes, or makes a costly decision), the question "who owns this?" becomes genuinely difficult to answer. Was it the data scientists who built it? The engineering team that deployed it? The business stakeholders who set the success metrics? The person who defined what "fair" means? Everyone can point to a piece of the system they don't own, and nobody has to own the failure.

I've heard CFOs push back on AI initiatives whose technology they never doubted, because they couldn't see the accountability chain. That's not caution. That's competence. They're asking exactly the right question, just about the wrong structure. They have spent their careers learning that a decision without a clear owner is a decision waiting to go wrong, and they can smell one across a conference table.

Your Org Chart Is a Lie About How AI Gets Made

Pull out the structure your company publishes. Find the AI or data science team. Trace the lines. It probably reports to a VP of Engineering, or the CTO, or maybe a Chief Data Officer. Clean lines. Clear hierarchy.

Now describe what actually happens when someone builds an AI system in your organization.

The data science team trains the model. But they didn't pick the business metric it optimizes for. Someone in product did, or maybe operations. They didn't decide whether to deploy it; that's usually a product call, sometimes influenced by finance.

They're not monitoring what it does in production. Operations handles that, or risk management, or sometimes nobody specifically. And they certainly aren't the ones making the call if something goes wrong.

So when the board asks "who owns this AI?" and you point to the org chart, you're not lying. You're just describing something that has nothing to do with how the decision actually gets made. That gap, between formal ownership on paper and actual accountability in practice, is where most AI governance breaks down.

It helps to be precise about two words we tend to use interchangeably. *Ownership*, in the traditional sense, means whose budget, whose performance review, whose headcount. On an org chart, that's usually clear. *Accountability* means something harder: you made the decision to deploy this, you understood the risks, and if something goes wrong, you face the consequences. You are the person who cannot pass the problem to someone else.

In most organizations, AI systems have plenty of ownership and almost no accountability. The data science VP owns the team. The product leader who defined what "good" meant owns the business decision. The ops leader who deployed it owns the infrastructure. The compliance team owns the audit trail. Finance owns the budget impact. Everyone owns a piece. Nobody owns the whole thing.

What that produces after a failure is entirely predictable. It's "well, the data was the problem," or "the business case was wrong," or "we didn't monitor properly." Each statement is true. Each points at a piece owned by someone else. And the failure floats, unclaimed, until it settles on whoever is least able to duck — usually whoever had their hands on the system last. Chapter 4 is about that dynamic and how to build accountability that survives it. The structural point here is narrower: nobody had to own the failure, because nobody was ever handed the whole thing to own.

Compare this to how an ordinary business decision gets made. If a CFO approves a capital expenditure and it goes bad, they own it. They had the authority to say no. They understood what could go wrong. They face the consequences. That structure (clear authority, clear accountability, clear consequence) is what actually drives good decision-making. AI hasn't broken accountability as a concept. It has simply exposed that most organizations never gave anyone accountability for AI decisions in the first place. They distributed ownership across multiple teams and called it governance.

Why the Org Chart Fails for AI

The problem starts with how data science evolved in most enterprises. Early on it was an analytics function: small teams reporting to an engineer or a data leader somewhere under finance, answering specific questions. Then AI became strategic. Suddenly these teams were making decisions that could reshape customer experiences, cost millions of dollars, or create regulatory exposure. But the organizational position didn't change. The team still reported to the same place, still had the same scope on paper, while making decisions several orders of magnitude larger than the ones the box on the chart was designed for.

Meanwhile, the functions with real authority weren't built to own AI decisions either. Product teams decide what features to build, not how to train models. Operations runs infrastructure, not model performance. Compliance audits outcomes, not systems before they ship. So you arrive at a structural trap: the people technically closest to the decision (data science) don't have the authority or the business perspective to own it, and the people who have the authority (business-unit leaders) aren't positioned to understand the technical implications.

This is why "let's build an AI governance committee" doesn't actually solve it. Committees are wonderful for coordination and terrible for accountability. Everyone on the committee has other priorities and other people they answer to. Decisions get diffused across the table until no single person is exposed. The org chart assumes ownership flows upward, that one person in the hierarchy is ultimately responsible. AI breaks that assumption by making the decision genuinely horizontal: it needs product perspective, technical capability, risk understanding, and compliance knowledge simultaneously, and no single person in the existing structure has all four.

Get this wrong and whatever you build will sit beside the real decision rather than on top of it. The reviews get scheduled; the actual call gets made somewhere else. A model ships behind a feature flag, so nobody has to call it a deployment. A policy gets "interpreted" for a pilot, and the pilot runs for two years. An approval happens in a corridor and gets backfilled into a ticket the following week. Nobody involved is cutting corners; the organization is routing around a structure that never touched the decision point, which is what organizations do, without announcement and without malice. Chapter 2 takes that failure mode apart in full. What matters here is the cause: the structure was built to record decisions, not to locate the person making them.

Three patterns show up again and again. The first is **accountability without authority**: risk and compliance teams are asked to govern AI but don't hold the budget, the technical fluency, or the position in the decision flow to prevent a bad deployment. They review after the fact. They are auditors, not governors. The second is **speed pressure meeting governance design**: "we need to govern responsibly, but we can't slow down," so you design something that looks rigorous on a slide, adds three weeks to every deployment, and still contains no single point at which a human being can say no. Teams under delivery pressure then do the rational thing and stop bringing you decisions: the small things go out as "experiments," and the large things get scoped and renamed until they qualify as small things. You end up faster *and* less governed: the worst of both. The third is **confusing process with governance**: the belief that more workflow equals better control. It doesn't. Better governance is clearer accountability, better information reaching the decider, and a real authority to say no.

The Three Risks in the Room

There's a scene that plays out in enterprises constantly, and it is not a joke even though it sounds like the setup for one: a CFO, a VP of AI, and a Compliance Officer walk into a decision. It usually ends the same way — with nobody entirely clear about who made the call.

Watch what each person is actually doing. The CFO's concern is straightforward: if this fails, how much does it cost us? They're thinking about downside, capital allocation, what they'd have to tell shareholders. **Economic risk** is their domain. The VP of AI is thinking about capability and execution: Can we build this? Will it work? What will we learn? They want to move, to stay competitive, to prove value. **Execution risk** is their domain. The Compliance Officer is thinking about exposure: What if a regulator asks? What if a customer complains? Can we defend this? They want documentation and an audit trail. **Regulatory and procedural risk** is their domain.

These aren't contradictory perspectives. But they are different perspectives, and they weight risk differently. The trouble is that your governance structure probably treats them as competing priorities to be reconciled, puts all three in the same meeting, assumes they'll reach consensus, and then acts surprised when they don't. Because nobody in that room is explicitly accountable for integrating the three views and making the actual decision. The CFO can't decide alone: they don't understand execution risk. The VP can't decide alone: they're not responsible for financial outcomes. The Compliance Officer can't decide alone: they're not responsible for business outcomes. Three people, none of them empowered to make the call.

So the meeting produces the most dangerous outcome of all: consensus that masks disagreement. You leave with an implicit agreement that "we're moving forward." But the CFO is thinking *we're moving forward if the economic model holds*, the VP is thinking *we're moving forward and we'll figure out the details*, and the Compliance Officer is thinking *we're moving forward and we'll handle the risks*. Those sound like agreement. They aren't. They're three different decisions that will collide the instant execution demands a real trade-off.

Put numbers on it. The system will cost $500K to build and maintain, and save $1.5M a year if it performs as modeled. The CFO says, "If it underperforms by 20% we still have positive ROI, so I can support this." Agreement subject to financial performance. The VP says, "I can build this, and I believe it'll work well." Agreement subject to being able to execute. The Compliance Officer says, "The financial case looks sound; I just need to see the governance around it." Agreement subject to governance being in place. All three have "agreed." Nobody wrote down the conditions. Then execution reveals that governance is murkier than Compliance wanted, or the build is harder than the VP expected, or the financial model was optimistic — and a decision everyone thought was made comes apart under the stress of reality. You never resolved the disagreement. You deferred it.

What Actually Works

The organizations that navigate this well share a structure that acknowledges the complexity instead of hiding from it. None of it requires new technology. All of it requires being explicit about things most organizations leave implicit, which is why it feels almost bureaucratic. It is. Do it anyway.

First, assign the decision to one person, not a committee. One name. That person needs the authority to say no to a deployment, to pull a system that isn't working, to redirect resources. And it should almost always be a business leader, someone whose P&L or customer outcomes are actually affected: a VP of Product, a line-of-business head, a COO. Because the decision being made isn't "is this technically possible." It's "should we take this risk, given our business situation." That is a business judgment, and it belongs to someone the cost will land on if it goes wrong.

Second, feed that person perspective, not consensus, at decision time. The decider needs to know what the model does, what can go wrong, what the business case is, what the regulatory exposure is, and what monitoring will reveal if it degrades — *before* they approve, not in a report six months later. Each function contributes its domain, framed as a decision input rather than a status update. The CFO lays out the financial envelope: "Here's the exposure if the model underperforms by 10%, 20%, 30%." The VP of AI lays out execution: "Here's our confidence in building this, here's what could break technically, here's how we'd mitigate it." The Compliance Officer lays out governance: "Here's what we'd need to document, here's what a regulator might ask, here's what we can defend." The decider's job is not "do we all agree?" It's "given what we now know, should we move forward?"

Third, document the decision explicitly. Not "we decided to deploy." Rather: "We decided to deploy this system, accepting these financial risks, with this mitigation plan, having documented this governance, knowing these specific scenarios could occur and here is how we'll handle each." That decision document becomes the reference point when reality shifts. If something goes wrong later, you can point to precisely what you understood and accepted at the time. That is the difference between a defensible decision and a scramble.

Fourth, give that same person real authority over what happens after. This is where most organizations fail without noticing. The decider approves; then operations runs it, risk monitors it, finance tracks cost, product owns the feature; and no one holds the authority to actually change course based on what's learned. You need the decision-maker, or someone they explicitly delegate to, empowered to reconfigure, halt, or redirect on the strength of post-deployment information, with no fresh trip through the hierarchy required. Otherwise the decision authority was theatrical from the start.

Here is what that looks like when it's real. You're deploying an AI system for credit decisions. The CFO sets the financial rule: "If this hits an 85% approval rate at a 4% default rate, I can fund it. If default goes above 5%, we recalibrate or stop." The VP of Risk, who owns credit, sets the governance rule: "I can defend these decisions only if I can explain them. If a regulator asks why we denied someone and we're running a black box, I'm exposed. I need model documentation and decision-level traceability."

The Compliance Officer sets the regulatory rule: “Fair lending is the focus. We need to show our approval rate isn’t substantially different across demographic groups, or, if it is, explain the difference through credit-risk factors, not demographic ones.”

Then the decision-maker integrates all three into something specific: “We’re deploying. Target is 85% approval at 4% default. We’re committing to documented, explainable decisions. We’re targeting demographic parity adjusted for credit risk. If either of the first two is violated, we recalibrate. If the third is violated, we stop. Person A monitors approval and default weekly; person B monitors documentation and explainability; person C monitors demographic parity. If any metric moves out of range, that person escalates to me for a decision.” That is a decision, one that integrates different perspectives and lays out what happens next. Not a consensus that hides disagreement.

The Boring Places It Breaks

If the structure above is where governance succeeds, the failures are almost always small, procedural, and easy to overlook. This is the heart of the whole discipline: AI governance is not won in the strategy offsite. It is won or lost in five mundane moments.

There’s **the decision meeting that decides nothing.** Three perspectives, no explicit authority, everyone talks until they’re tired, and someone says, “Okay, sounds like we’re aligned on moving forward.” You’re not aligned. You’re tired. Fix it by writing down what was agreed, documenting the conditions and assumptions, and having the decision-maker say plainly: “Here’s what I’m deciding, and here’s what I’m accepting by deciding it.”

There’s **the framework that exists but gates nothing.** Someone has to make a decision *from* the checklist: not “is it complete?” but “given what it shows, should we deploy?” If the answer on the checklist is always yes, you have a form, not governance.

There’s **the risk discussion that’s all upside.** Everyone’s talking about the value this will create; nobody’s talking about what happens if it doesn’t. If the CFO isn’t laying out financial downside, the VP isn’t laying out execution risk, and Compliance isn’t laying out regulatory risk, you haven’t assessed risk. You’ve just decided you like the idea.

There's **the monitoring plan nobody monitors.** You'll "track accuracy" and "watch for bias," and you've handed it to a team that has other priorities. Months pass. The first time anyone looks closely is the first time something breaks. Replace "the data science team will keep an eye on it" with "person X checks this metric weekly, and person Y decides what to do if it moves."

And there's **the escalation that doesn't escalate.** Something unexpected happens, the monitoring catches it, an email goes out — and it gets lost, or lands with someone who can't act, or reaches a committee that meets quarterly. By then it's late. Build a path where an observation actually reaches someone with the authority to act, fast enough to matter.

None of that is exciting. All of it is the job.

What a Decision Sounds Like

Sooner or later somebody asks why you deployed the system. It might be a director, a regulator, a plaintiff's lawyer, or your own successor reading the file two years after you've moved on. There is a version of the answer that ends the conversation and a version that starts an investigation. The good one sounds like this: "We evaluated the deployment against financial risk, execution risk, and governance risk. The financial model showed positive ROI with acceptable downside. We had the technical capability to execute. We had a governance and regulatory framework we could defend. We assigned clear ownership and a clear escalation path. We made the decision deliberately, and we documented what we understood and accepted."

That is what a governance decision sounds like. Not "everyone agreed." Not "the model was validated." Not "it was on the roadmap." But: we thought it through, we integrated perspectives from across the organization, and we were willing to own the consequences. What a board should then *do* with an answer like that (the oversight cadence, the reporting it should demand, the risk appetite it ought to be signing) is Chapter 5's territory. The prerequisite is having an answer worth overseeing.

When you have that clarity (about who decides, what they're deciding, what they're accepting, and what happens if things change), the CFO, the VP, and the Compliance Officer stop needing to reach a consensus they were never actually going to reach. Each provides a clear view of their own domain. One person integrates those views into a decision. And everyone knows what was decided and why. Ask "who owns this

AI?" and the answer is one name. Not a team. Not a committee. One person who made the call and can defend it. That clarity, about who decided and why, is what governance actually looks like. Everything else is just process.

FIGURE 1.1

The AI Decision Accountability Map

Worked example: a credit-limit model at a regional lender.

LIFECYCLE STAGE	R RESPONSIBLE	A ACCOUNTABLE	C CONSULTED	I INFORMED
Objective why we're building it	Product lead	**BU head**	Finance	Exec team
Data what it learns from	Data eng	**Data owner**	Privacy	Product
Model how it decides	ML team	**ML lead**	Risk	Product
Deploy who it reaches	Platform eng	**No one** owns deploy	Security	BU head
Monitor what counts as wrong	MLOps	**Risk lead** **Ops lead**	Data eng	Audit
Respond who acts when it fails	Support	**BU head**	Legal	Customers

One A per row. No blanks. No ties.

The Model row is the shape to copy: one name in A, and only one.
Deploy has nobody there. Monitor has two — in an incident, that is the same thing.

Workshop use: a cross-functional team fills one row per stage for one live system.
Every blank or doubled A becomes a named owner before anyone leaves the room.

FIGURE 1.1

The Bottom Line

Your AI strategy became a governance problem the moment a single model's behavior started depending on data, modeling, product, operations, and risk simultaneously. Five domains your existing structure assumed one person owned. That is why "who approved this?" stops the room. There is no shortage of ownership; there is no accountability, and the meetings end in a consensus that hides three different unstated conditions. More process does not fix this, because more process is usually the symptom. The fix is an organizational design choice you can make this quarter without buying anything: put one business leader in the decision seat, hand them perspective instead of agreement before they decide, write down what they accepted and what would change their mind, and leave them standing authority to halt or redirect once the system is live. Do that and the room has an answer. Do less and you have a chain of custody with nobody at the end of it.

Field Kit

1. The five questions to bring to any AI deployment decision. Do not leave the room until each has a specific answer with a name attached: - Who has the authority to say no to this deployment — one person, not a committee? - What exactly is that person accepting by approving it (financial downside, execution risk, regulatory exposure)? - What are the specific thresholds that trigger a recalibrate, and which trigger a full stop? - Who monitors each of those thresholds, how often, and who decides what to do when one moves? - How does an alarming observation reach someone with authority to act — and how fast?

2. The ownership-vs-accountability audit (per system). For a given AI system, list every function that touches it (data, modeling, product, operations, risk, compliance, finance) and write down what each *owns*. Then ask the separate question: who is *accountable* — who faces the consequences and cannot hand off the problem? If that second column has more than one name, or none, you have diffusion, not governance.

3. Diagnostic: signs your accountability is distributed, not clear. You likely have a problem if: your data science team and your AI-ethics/compliance team sit in different chains of command and don't coordinate before deployment; your governance committee reviews projects but cannot block a deployment; you use three different job titles for people doing the same work; the first five conversations after a failure are about whose fault it was, not how to prevent a repeat; two leaders make conflicting calls on the same system because they optimize different metrics; or risk and compliance are most engaged *after* deployment rather than before.

4. The decision-record template. For every deployment, capture in one page: the decision-maker (one name); the decision ("we are deploying X"); what is being accepted (financial, execution, regulatory risks, stated explicitly); the mitigation plan; the specific recalibrate-and-stop thresholds; the named monitors and cadence for each; the escalation path; and the date. This is the artifact you point to when the board — or a regulator — asks why.

5. The three decision points every AI system needs an owner for. Assign a name to each, and confirm each owner has the authority the role requires: - *Deployment approval* — who says "yes, this goes to production," positioned so the cost lands on them if it goes wrong. - *Failure monitoring* — who watches for the failures that matter (bias, drift, decision-quality decay, not just downtime) and has the authority to pull the cord. - *Objective definition* — who decides what the model is actually supposed to optimize for, translating business intent into training-time constraints, so a technically accurate model isn't a business mistake.

CHAPTER 2

Governance by Design, Not Governance Theater

Part I · Why AI Is a Governance Problem, Not a Technology Problem

I once walked into a Fortune 500 company's AI governance meeting and saw the org chart before I saw a single decision. Fourteen people: Legal, Compliance, Risk, Product, Engineering, Finance, and three layers of leadership. The program had been running for six months. It had produced a great deal of documentation and exactly zero shipping decisions. When I asked the head of the program what problem the committee was actually solving, she said, without irony, "We're ensuring governance."

That's not governance. That's the appearance of governance.

Governance is a decision-making framework. Chapter 1 made the case that a committee cannot be the accountable party, and I'll take that as settled. What that room was missing was narrower and more fixable: nobody had specified which people needed to be involved at which decision points, with what authority, and with what consequence for being wrong. Fourteen people in a room is not a governance structure. It's an attendance list, and the difference between the two is the difference between being in the room and being the one who decides.

This chapter is about that difference, and about a claim that sounds underwhelming until you've watched a few of these programs fail: real governance is not a committee, a checklist, or a paper framework. It's a small set of boring, load-bearing decisions embedded in how the work actually happens. The glamorous version (the ethics board, the five-layer review, the framed value statement) is where governance goes to hide.

Here is the reframe this chapter is built on.

The myth: AI governance means standing up a committee, a checklist, and a multi-layer review framework, and you have governance once those exist.

The reality: real governance is a small set of boring, load-bearing decisions embedded in how the work actually happens, each with one named person who decides and owns the outcome.

How committees stop deciding

Watch what happens when you design governance through consensus. Finance wants to minimize cost, Legal wants comprehensive documentation, Product wants speed, Engineering wants stability. And they all get a vote, so the only outcome that keeps everyone happy is the one that offends everyone equally. Everyone signs off. Nobody is happy. And (Chapter 1's point, so I'll leave it at one line) nobody is accountable either. I've watched this at companies full of smart people making individually rational choices. Nobody sets out to build a committee that decides nothing. They're following the logic of "if we want everyone to agree, we need to make decisions nobody disagrees with," which resolves to decisions that satisfy nobody.

What I want to add is the question that almost never gets asked *before* the committee is chartered: what are we actually trying to prevent?

The answers are not interchangeable, and they don't want the same structure. Preventing *systems that break the law* means one narrow compliance rule and someone with authority to enforce it — a veto, and a small one. Preventing *systems that harm customers* means product standards with a named person accountable for meeting them. Preventing *systems that lose money* means cost constraints owned by Finance, with actual numbers attached. Preventing *systems that break architecturally* means technical standards and an engineering owner who can enforce them. Preventing *systems that embarrass you* is a policy problem, not a governance problem, and confusing the two is why your model review somehow has an opinion about tone of voice.

Five problems, five mechanisms, five owners. Not one of them is a committee. A committee is what you get when the answer is "all of the above, plus whatever we haven't thought of yet." That is the answer when the prevention conversation never happened, so the structure has to cover every conceivable scenario at once. Write rules for every scenario and nothing ships, and the governance structure becomes the reason you can't build AI at all. Which is, technically, an outcome. Just not a governed one.

Here is a sentence I have never once seen in an organizational document: *"The AI Governance Committee will not make decisions. Its purpose is to flag constraints and escalate conflicts. Decision authority rests with [specific role]."* That one sentence would dissolve most governance theater on contact. The theater survives precisely because nobody wants to say it aloud in a room where everyone was invited.

The deeper trap is that AI governance is not primarily a legal or compliance problem. It's a speed-and-risk trade-off, and committees are built to minimize disagreement: a different objective from moving fast, and a different one again from managing risk. So you get slower decisions that are also less careful, because by the time a proposal reaches the group it has been ground down so long that nobody in the room is really reading it.

The framework that was a process without a decision

The natural response, once a company realizes its committee decides nothing, is to build something more rigorous. More steps. More reviews. More sign-offs. This is where the second, subtler failure lives, and it's worth dwelling on, because it fools people who have done the work.

A company I worked with built an AI governance framework with five layers of review. Data quality: is the training data sound? Model validation: does it perform as expected? Fairness: does it produce disparate impact? Business impact: does deploying this make sense? Executive approval: final sign-off. On paper, comprehensive. In practice, close to useless.

A model came through. The data-quality team reviewed it and approved it. The validation team checked the metrics and approved it. The fairness team looked at the aggregate statistics and approved it. The business team saw revenue potential and approved it. The executive signed off. The model deployed. Within months it was systematically making worse decisions for a specific population segment — the kind of thing everyone in the room would have called unacceptable if you'd described it in advance.

The conversation afterward went almost exactly like this:

> *"The fairness team reviewed this." "Yes, we looked at the aggregate statistics." "Did you notice the disparate outcomes?" "No, the aggregate statistics looked fine." "The model is producing disparate outcomes." "Then the environment changed, or we made a wrong assumption." "Your approval was supposed to prevent this." "Our approval was that we followed the process."*

Read that last line twice. The framework had done exactly what it was designed to do: document that a process existed and distribute responsibility across five parties. What it had not done was prevent a bad outcome, because at no point did anyone have to say, "I understand all of these reviews, and I am choosing to accept the risk of deploying this model given these constraints." Each reviewer answered their own narrow question and passed it along. Approve, approve, approve, approve, approve.

It was a process without a decision. Everyone did their part. The system reached "approved." Nobody actually decided.

Alignment is not accountability

The most expensive version of this failure I know involved a financial services company and a model built to assist with loan decisions. It had been through the full responsible-AI process. Bias testing: passed. Fairness audit: passed. Explainability review: passed. Human review: passed. Every box checked, before deployment, by competent people.

Six months into production, regulators examined the loan decisions and found a disparate-impact pattern: the model was making different decisions for similar applications along protected characteristics. Not intentionally. It had learned the pattern from historical data. The company pulled the model and investigated, and the investigation surfaced the truly uncomfortable part: the bias had been detectable in the *test* data. One of their own data scientists had flagged it in a meeting. But because no single person was explicitly responsible for deciding "is this level of bias acceptable for this use case?", the concern got absorbed into the "we'll monitor that" pile and forgotten. Alignment checkboxes: all satisfied. Accountability: nowhere. The result was a significant regulatory penalty and real legal liability.

Here is the distinction those checkboxes obscured. A responsible-AI checklist is about *alignment*: does this model align with our values, meet our standards, reflect us doing the responsible thing? Alignment means you've thought about the right things.

Accountability means someone is responsible for the outcome if that thinking turns out to be wrong. They are not the same, and the gap between them is where the liability lives.

Your framework says “we will test for bias,” but not who is responsible if bias slips through. It says “we will have human review,” but not what happens if the reviewer misses something. It says “we will monitor for drift,” but not who owns the decision to pull the model when drift shows up. Those gaps stay invisible right up until the day someone is harmed and your legal team needs to draw a straight line from the harm back to a specific decision by a specific person with clear authority.

This is also why your lawyers care about accountability far more than they care about your checklist. Liability doesn’t come from failing to be responsible. It comes from failing to be clear about who was responsible. If you can show that you understood a risk, that someone with authority made a judgment about accepting it, that the judgment was documented with reasoning, and that it was approved at an appropriate level, you have a defensible position even when the outcome is bad: you made a reasonable decision with the information you had. If you can’t show those things, then a passed bias test reads, in a courtroom, as evidence that you saw the risk and didn’t take it seriously. Your checklist is evidence. It was never a shield.

What governance by design actually looks like

If committees are the appearance of governance and multi-layer frameworks are process without decision, what does the real thing look like? It looks smaller and more boring than any of it. The companies handling this well have made a structural choice: they split the decision into parts, and they are ruthless about which part is a vote and which part is not. Constraints are not votes. That single distinction does more work than any charter document I have ever read.

Who decides what gets built. Product and Engineering. Speed matters here, so the committee has no role. They operate under constraints (legal, compliance, risk), but again: constraints are not votes.

Who decides whether the constraints are met. Legal, Compliance, Risk. Their job is not to vote on the idea; it’s to flag when a proposal violates a constraint. If the constraint is “all customer data must be encrypted in transit,” a proposal that doesn’t gets sent back. The builder then either changes the proposal or moves to change the constraint, and changing the constraint escalates.

Who has authority to change the constraints. The CFO and executive leadership. This is the *only* place consensus belongs, because changing a constraint means changing the risk appetite or the business model. If the rule is "all AI systems must use approved models" and Product wants an unapproved one, that's not a product decision — it's a leadership conversation about risk.

Who owns the outcome. Product and Engineering, with escalation to leadership when constraints are violated. This is the load-bearing piece: the people who make the call own the consequences. If they ship something that breaks compliance, it's on them, not on "the committee."

Notice what this structure buys you. Decisions move fast, because there's no voting. It's explicit about what is being constrained and why, and unambiguous about who is accountable. And, counterintuitively, it manages risk *better* than consensus does: a named team that knows constraint violations get flagged, and knows those flags escalate to someone with real authority, thinks harder about what it ships. This is not "let Engineering do whatever it wants." It's "Engineering makes the call, inside clearly defined constraints, with escalation when those constraints conflict with a business goal." Which is also fairer than a democratic committee that never decides anything.

But separation of concerns isn't enough on its own, because you can draw that org chart and still end up with the five-layer failure. What the good structures add is the thing the framework was missing: a decision layer. Somewhere in the process, a person (or a genuinely small group, not a committee) has explicit authority to decide *should we use this model?* They don't validate every detail; they rely on expert input. But they synthesize it, make the choice, and write it down in the decision record from Chapter 1.

Add one field to that record. After the constraints, after the thresholds, after the named monitors, the decision-maker writes the sentence that governance by design actually turns on:

> *Here is what would make me reverse this decision.*

It reads like a formality. It isn't: it is the only field in the record its author has to write against their own interest. Everything else documents why the decision was right; this one documents, in advance and in their own words, the conditions under which it stops being right, drafted at the exact moment they have every incentive to believe it will all work out. That is uncomfortable to write, which is why most decision records don't have it, and it is what changes the incentive. The question when the model misbehaves stops being "did we follow the process?" (which has no useful answer) and becomes "we said X would reverse this; has X happened?" A decision-maker who has to name their own reversal conditions stops optimizing for a complete checklist and starts imagining what could actually go wrong.

You could write this whole approach on one page. You don't need a framework at all. Organizations build one anyway because nobody believes "we had a thoughtful person decide" will survive an audit, so they build the five-layer review for audit-readiness and mistake it for risk management. The best outcomes come from running both deliberately: a framework, for the auditors, *and* a decision, for the actual risk. If you already have a comprehensive framework (and you probably do), don't tear it out. Add the decision layer on top, and make sure whoever holds it knows their job is to make a defensible decision, not to certify that everyone else did their part.

The boring org chart that makes it real

At this point the instinct in most executive meetings is to hire the problem away. "We need an AI governance lead." Or an AI ethics officer. Or a Head of Responsible AI. You post the role, get four hundred applicants, and hire someone smart and genuinely committed. Six months later you discover the role doesn't work. The problem isn't the person; it's that you've put one individual in charge of something that requires organizational alignment, and alignment is almost never something one person can conjure alone.

I watched exactly this play out at an organization that hired an excellent AI ethics leader with a strong mandate and a reporting line to the CHRO. It looked good on paper. Then a deployed model began degrading in one customer segment, and nobody knew who was supposed to respond. The data science team thought it was the business leader's problem: their use case, their responsibility. The business leader thought it was data science's problem: they built and deployed it, they should fix it. The ethics leader tried to coordinate and ran into the limit of her job: she could convene both teams, write up the issue, escalate it, put it on a slide. She could not make either team move. A mandate is not authority. She had been hired to hold responsibility for a system she had no power to change, which is not a role — it's a liability with a title. Ambiguity is where AI governance goes to die, and that is the specific ambiguity that kills it.

The organizations that actually enable governance aren't the ones with the best governance officer. They're the ones with the clearest accountability structure. The question was never "how do we hire AI talent?" It's "who owns the decision?" And the answer comes down to three accountabilities you make explicit and assign to people who already exist in your business.

Model ownership. Not data science, not IT — actual business ownership. Someone who owns the business outcome of that model, understands when it's breaking, and is accountable for the choice to keep it running or pull it. They don't need every technical detail; they need to connect technical signals to business impact, and they need the authority to say "pull it" and the responsibility that comes with choosing not to.

Data accountability. Someone owns what data is in the system: where it came from, whether it's been validated, what happens when it changes. This is distinct from model ownership: you can have a perfect model running on garbage data. "IT owns data" is too vague. Someone specific owns *this* model's data, its lineage, and its fitness for the decision being made.

Governance authority. Someone has explicit authority to pause or pull a system when requirements aren't being met. Not advisory authority. Not "we should probably review this." Authority. If the data owner flags degraded quality or the model owner flags bad performance, governance says "pause until this is resolved." And the pause actually happens. This can't be a person without organizational power; it has to be someone whose decision gets implemented.

None of this is glamorous. You're not hiring ethics officers with impressive titles; you're clarifying who is responsible for what, using people who already own parts of the system. The model owner comes from the business, the data owner from engineering or analytics, the governance authority from risk, compliance, or operations. They aren't new roles; they're explicit accountability. Crucially, they have to be able to decide together, not through a committee, but through clarity. The model owner says, "based on the business outcome, I want to change how we use this." The data owner says, “we can’t change it without understanding these data implications.” The governance authority says, “I need this much visibility before anything changes.” And then they solve the problem, because each dimension of it has an owner.

That’s the boring reality: AI governance isn’t a function. It’s a structural property. It emerges the moment you’ve clarified who decides, and it evaporates the moment you haven’t. Your best ethics officer can’t create it alone; clear accountability enables it even with people still learning the domain. We like to believe good governance comes from good people. More often it comes from clear roles. The name on the door matters far less than the clarity behind it.

Governance doesn’t end at deployment

There’s a final myth to kill, and it’s the one that makes everything above feel finished when it isn’t. In the conference talk, “responsible AI” sounds like something you’d frame on a wall: a commitment, a value, a north star. In production, it looks like a spreadsheet. A checklist nobody finds interesting. Someone logging into a dashboard at 3 a.m. because an anomaly flagged, and they need to know whether the model just broke or whether the data shifted underneath it.

That’s not the story we tell. We tell stories about bias-detection frameworks, fairness metrics, ethics review boards: all real, all important, and none of them the thing that makes the difference in organizations actually running AI in production. I once talked with an engineering leader at a financial services firm who’d built what looked like an exemplary responsible-AI program (fairness audits, bias-mitigation protocols, a genuinely rigorous review process) and who kept discovering problems anyway. Not because the governance framework was catching them. Because the *monitoring* was good enough to catch something changing, and then the team had to work out what changed and why. That visibility was what made them responsible.

Here is the distinction that matters more than any checklist. Responsible AI *in theory* is about making a good decision about whether to deploy. Responsible AI *in practice* is about having enough visibility into what the deployed model is doing that you can react to it. Deploying is the beginning of the work, not the end: it’s when you find out whether your fairness metrics predicted real behavior, whether your bias mitigation touched the biases that matter in your context, and which edge cases your testing never caught.

Building that visibility is a discipline in its own right (which signals matter, what to log, how to tell a healthy system from a merely healthy-*looking* one), and Chapter 12 is where that work lives. The governance question is smaller and comes first: when one of those signals moves, whose decision does it land on? A dashboard with no route back to a decision-maker is an expensive way to document your own decline. And the number on it is rarely the point anyway; 87% hides more than it reveals.

Which brings me to the failure I see most often in organizations that got everything else right. A decision record is a dated document: given what I knew on this date, I accepted these risks under these constraints, and here is what would reverse it. Then eighteen months pass. The person who signed it has moved to another business unit. Two of the reversal triggers were never instrumented, because nobody who could have wired them up was in the room when they were written. And the use case has widened: the model that scored one product line now scores three, and nobody re-decided anything, because nobody experienced it as a decision. None of that is a monitoring failure. All of it is a governance failure, invisible until someone asks who currently owns the call to keep this thing running and the honest answer is that nobody has since the reorg.

The fix is as boring as it sounds: every decision record carries a review date and a named successor alongside its reversal triggers, and every trigger has someone who has actually instrumented it, because an uninstrumented trigger is a wish. That's the whole mechanism — two lines nobody will ever congratulate you for maintaining.

Oversight is saying "no" before launch. Visibility is saying "here's what's actually happening" after launch. Both matter, but visibility only *governs* anything if a divergence escalates to the person who owns the decision and can reverse it. You don't finish responsible AI. You keep running it. That spreadsheet nobody finds impressive is where it actually lives.

The Bottom Line

Committees, checklists, and multi-layer frameworks feel like governance because they produce artifacts, but their real function is to distribute responsibility until no one holds it. Before you charter a committee, ask what you are actually trying to prevent; every honest answer wants a narrow mechanism with a named owner, not one body trying to prevent everything. Real governance is smaller and duller: separate the decision into who builds, who flags constraints, who can change constraints, and who owns the outcome (because constraints are not votes), then make one named person with authority actually decide "we deploy this, given these constraints," and write down what would reverse it. Wrap that decision in three explicit accountabilities: a model owner, a data owner, and a governance authority who can actually pause a system. And keep the decision alive after launch with a review date and a named successor, because a decision nobody revisits is just a dated document. The test is brutally simple: if you can't answer "who decides whether we build this, and who owns it if it breaks?" in one sentence, you don't have governance. You have the paperwork governance leaves behind.

Field Kit

1. The one-sentence test. Before any AI governance structure is called "done," a leader must answer both aloud: *"Who decides whether we build this?"* and *"If it breaks, who owns fixing it — and if we disagree about what 'broken' means, who decides?"* If either answer takes more than a sentence, you have theater.

2. The prevention conversation (have it *before* you charter anything). Name what you are actually trying to prevent, and give each answer its own mechanism and owner: breaking the law (narrow compliance rule, enforcement authority) · harming customers (product standards, accountable product owner) · losing money (Finance-owned cost constraints, with numbers) · embarrassing you (policy, not governance) · breaking architecturally (technical standards, engineering owner). If the answer is "all of them," you are about to build a committee.

3. The four-part decision split. Assign each in writing, and mark which are votes: - *Who decides what gets built?* → Product/Engineering (not a vote) - *Who flags whether constraints are met?* → Legal/Compliance/Risk (not a vote — a flag that can escalate) - *Who can change a constraint?* → CFO/executive leadership (the only place consensus belongs) - *Who owns the outcome?* → the builder, with escalation on violation

4. The reversal clause. Chapter 1's decision record is the artifact; this is the one field to add. Before any model deploys, the accountable person writes, in their own words, the conditions under which they would reverse the decision — named metrics, named thresholds, named changes in the use case. Not "if performance degrades." Something you could check on a Tuesday and get a yes or a no. If a decision-maker can't name what would change their mind, they haven't decided; they've expressed a preference.

5. The three-owner grid. For each production model, name a person (not a team) in each row: **Model owner** (business outcome; authority to pull) · **Data owner** (lineage, quality, fitness; authority to flag) · **Governance authority** (authority to pause until resolved). Any blank cell is a live governance gap.

6. Alignment-to-accountability upgrade. Audit your responsible-AI checklist. For every line that says "we will test/audit/monitor for X," add: "*…and if X exceeds threshold [Y], [named person] decides whether to accept or remediate it, documented with reasoning, approved at [level].*" Turn every process step into a decision point with an owner.

7. The decision-expiry check. For every model in production: each reversal trigger is instrumented by a named someone (an uninstrumented trigger is a wish); a tripped trigger lands on a person who *still holds authority to act on it,* not their predecessor and not a queue; and the record carries a review date that hasn't passed. Then the question that catches the rest — has the use case widened since the decision was made? If so, your decision covers a system that no longer exists.

PART II

Decision Rights, Accountability & the Board

CHAPTER 3

Who Decides: Decision Rights at Scale

Part II · Decision Rights, Accountability & the Board

Two companies, the same week, opposite solutions to the same problem, both failing.

The first had a centralized AI team: every model reviewed by one group, one standard, one queue. The second had pushed everything to the business units: whoever owned the outcome owned the model. One was slow. One was incoherent. Neither could answer who approved this. Both leadership teams believed they'd made the responsible choice, and both were wrong in a way that took months to surface on a balance sheet.

The centralized company thought it was optimizing for control. What it got was a bottleneck. The distributed company thought it was optimizing for speed. What it got was risk. Neither understood the actual constraint: at enterprise scale you need speed and consistency at the same time, and the structure that gives you both has nothing to do with where you put the team. It has to do with where you put the decision.

Here is the reframe this chapter is built on.

The myth: governing AI at scale is a choice between centralizing the AI team for control or decentralizing it to the business units for speed.

The reality: speed and consistency come from where you put the decision, not where you put the team: authority belongs with whoever knows the problem, can move fast enough, and owns the consequences.

The question you've been arguing about (*should we centralize or decentralize AI?*) is the wrong one. The right question is: for each kind of decision, who has the authority to make it, can they make it fast enough, and do they own the consequences when they're wrong? Get that right and the centralize-or-decentralize debate dissolves, because the answer turns out to be "both, at different levels, deliberately."

Why both extremes fail

A centralized AI team sounds excellent in theory: one group of experts sets the standards and approves the decisions, and nothing ships that violates them. What actually happens is that the team becomes the thing every unit waits for. "We could solve this with AI, but it'll take six months to get approval," a unit says, and then either doesn't solve it or builds a quiet workaround. Worse, a central team can't hold every business context in its head. "We only use approved models." But different units have radically different requirements. "We monitor for fairness." But fairness means something different for hiring than for lending than for content recommendation. You don't get strong governance. You get slow governance that gets bypassed.

Decentralization fails the opposite way, and it feels great until it doesn't. Every unit moves fast and understands its own problem — and then the hiring team builds a resume screener, the recruiting team builds another, L&D builds a third, none of them talk, and each makes different choices about data, models, and monitoring. One uses a model biased against a protected class and never notices, because no one thought to check. Another breaks in production and loses customer data because nobody owned operations. A third builds a black box no regulator will accept. And here's what turns fragmentation from annoyance into existential threat: when something goes wrong in one unit, the regulatory exposure is enterprise-wide. The hiring team's bias problem is a *company* problem. Decentralization feels fast until the first serious failure, and then it's expensive in a way no amount of local velocity paid for.

Neither pattern holds, and scale shows why. In a startup with five models, centralization works fine: everyone knows the models, the CEO approves changes. In a growth-stage company with fifty, you decentralize; teams own their performance, you publish principles, most things work. Then you're an enterprise with five hundred models across fifty business units and multiple regulatory regimes, and both instincts collapse at once. Pure centralization stops scaling around fifty to a hundred projects: the bottleneck gets so obvious people build shadow processes to escape it. Pure decentralization stops working around twenty to thirty, as the inconsistency gets costly.

Read those numbers narrowly. They map one axis and one axis only (whether authority is pooled or distributed), and they say nothing about how much governance machinery a portfolio of that size needs, which is a separate question with its own thresholds and gets its own treatment later in this book. Nor are they a law. The useful claim is not that project fifty-one breaks centralization; it's that both structures have a ceiling, the ceilings are closer than anyone plans for, and you will hit one of them. You didn't choose wrong. You outgrew the choice.

Authority at the decision point, not the team location

So the real question is never "where does the team sit?" It's "where does the authority sit?" And the move that changes everything is to put authority at the decision point rather than the team's location.

In the centralized company, authority for every AI decision lives with the central team: a unit says "We want to build a hiring screener," and the central team decides. That's the failure, because the central team doesn't know enough about hiring to make that call well. In the decentralized company, authority lives with the unit, but the company can't hold a consistent line or manage enterprise risk. What works is splitting authority so each piece sits where the knowledge and the consequences are. The business unit decides, *"We have a problem and we think AI can solve it"*: they know their problem better than anyone. A shared standards function says, *"If you use AI this way, here's what you owe around data quality, fairness, explainability, and operational risk."* Constraints, not approvals. The business unit decides, *"We'll meet those constraints with approach X."* Again, their call. And a shared audit function checks, *"Are you actually meeting what you committed to?"* Not blocking. Verifying.

That distribution buys all four things the two extremes each sacrificed: speed (units aren't waiting for approval), consistency (everyone builds to the same constraints), risk management (you check for problems without blocking), and accountability (it's clear who decided what and why).

The whole thing turns on a subtle, load-bearing distinction: approval authority versus standard-setting authority. Approval means "we decide whether you can do this," which slows the organization down. Standards mean "here are the constraints; meet them and you're free to move," which speeds it up. A real standard reads like this: if you use AI for pricing, document your data sources and their known biases, verify the recommendations don't systematically harm specific customer segments, keep a manual override, and review outcomes quarterly. Meet that and you build without waiting (no gate), and six months later the standards team verifies you *actually* did it.

The centralized company does approval; the decentralized company ignores standards; neither does the thing that works.

This is where companies sabotage themselves, always the same way: they write standards that are secretly approvals. "All AI systems must use approved models" isn't a standard: it's a gate wearing a standard's clothes, and units will either ignore it or lose six months getting a model blessed. Real standards are verifiable statements of what good looks like: a documented rationale for data sources; outcome monitoring for consequential decisions; a named person accountable for outcomes; an escalation process for when a system drifts outside its training distribution. Those clarify. They don't gatekeep.

The hierarchy: three levels, three speeds

Splitting authority between "the unit" and "a shared function" is the first cut. The deeper structure is that AI decisions don't live at one altitude. And most organizations treat every decision as if it belongs at the same level. They stand up a governance board that decides which models to deploy, what risk thresholds are acceptable, how to handle fairness, what data can be used. Everything through one structure. It works until the day the board finds itself debating whether a model should log timestamps in a particular format, or whether a team can run a quick experiment. Now it's a bottleneck, not because it's doing anything wrong: the decision in front of it belongs somewhere else entirely.

Enterprise AI needs three levels of decision-making, each with its own questions and its own natural speed.

Enterprise level is strategy, architecture, and the constraints everyone lives inside. What's our risk appetite for AI? Which vendors? What data can be used? How much are we spending (a CFO-level call)? These affect multiple lines of business and set the boundaries others work within; they should move *slowly*, be made rarely, and belong to people who see across the whole organization. If you're revisiting enterprise AI strategy every month, something is broken.

Line-of-business level is which initiatives to pursue inside a business and how to implement them. Build a demand-forecast model? Buy this vendor's platform or build our own? These affect one area, not the whole strategy; they're guided by the enterprise constraints (spend this much, use this vendor, follow this risk appetite) but made locally. A leader should be able to pursue an initiative without a trip back to the board. If the constraints are clear, the local decision is already inside them. Move.

Team level is how to build, test, and deploy within an initiative. What features? Which algorithm? How do we handle missing values? Implementation choices, guided by the line-of-business decision but made by the people doing the work, and they should move *very* fast. A data scientist should be able to choose an algorithm without asking permission or waiting for a governance vote.

Organizations get this wrong in two diagnosable directions. They **push decisions up**: a team takes every experiment to the board until the board is a bottleneck and everyone's frustrated. That almost always means the higher levels never set clear enough constraints; the team asks permission because it doesn't know what it's allowed to do. The failure is upstream but shows up as slowness downstream. Or they **push decisions down**: teams pick vendors and set their own risk thresholds, and now one unit runs Vendor A tolerating 3% bias while another runs Vendor B demanding 0.5%, and none of it adds up to a strategy. Same root cause, opposite direction.

You can tell you've got it right by the feel of each level: enterprise decisions feel *weighty* ("what's our risk tolerance?" not "should we tune this hyperparameter?"), and team decisions happen without anyone asking. If your board's agenda feels trivial, you've pushed decisions up; if your data scientists need sign-off to run an experiment, you haven't pushed them down far enough. The organization that gets this right moves faster where it should and slower where it should, so the whole system is both faster and more coherent than one funneling everything through a single gate.

Structured coordination: decide by impact and reversibility

Three levels tell you the altitude of a decision. They don't tell you how much process any given decision deserves, and that's the gap where governance dies. You've heard the sound it makes. It happens in an architecture review, when someone asks, "Who decides whether this model gets deployed?" and the answer comes back: "Depends. If it's strategic, corporate. If it's tactical, the team. If it crosses business units, we need alignment. If it's in production, the data team has final say. If it's high-risk, maybe compliance." That is not a governance framework; that is a coordination problem nobody has solved, read aloud.

The insight that cuts through it: different decisions need different authority *based on impact and reversibility*, not on size or how strategic something sounds. A small model that seemed unimportant at birth can grow huge; a strategic-sounding project can be trivially reversible. So sort on those two axes. **Low-impact and reversible** (deployed to a small segment, easy to switch off): delegate to the team, notify the committee, no approval. **Medium-impact and partially reversible** (customer-facing, costs money to unwind): require approval, but streamlined. One designated reviewer, not a committee; a decision in forty-eight hours, not thirty days. **High-impact and irreversible** (high-stakes customer outcomes, regulatory exposure, major infrastructure): full committee, all stakeholders, the seriousness of a major capital investment. **Cross-boundary** (touching another team, geography, or shared infrastructure): a fast committee of only the people with real skin in the game.

The framework is simple; all the power is in the rigor, which is exactly what most organizations skip. They say "get approval from the team" without defining what approval means, how long it can take, or who holds a veto. Structured coordination demands four things written down and honored: **clear criteria** (what makes something "high-impact": customer reach, financial exposure, reversibility, compliance surface); **clear authority** at each tier (team lead for low; parallel technical and compliance reviewers on a forty-eight-hour clock for medium; a defined committee for high); **clear escalation** (if a team disagrees with a rejection, where does it go: to the CTO, the CRO, a business leader?); and **clear documentation**, not for show but so that in six months someone can read *why* the model was built this way, and the next team inherits a precedent instead of relitigating it.

All of that runs on discipline (people following the process instead of routing around it), and it costs real overhead: documentation time, forty-eight-hour waits, the occasional project delayed for an alignment nobody enjoyed. Price it into the model development timeline the way you price data cleanup, as a line item rather than a surprise. What you buy is that you stop making the same decision five times and getting five different answers; you stop discovering after the fact that someone shipped a thing violating a principle everyone had agreed to; and you stop having the "but we're different, those rules don't apply to us" conversation, because being different was never one of the criteria.

And you stop something more insidious: decision authority becoming a shadow power structure. Without it, the real authority lives with whoever argues best, holds the most political capital, or is most willing to route around the system — none of which has anything to do with making good AI decisions. Structured coordination drags authority back to criteria, harder to dismiss and easier to defend than charisma. The question was never "who decides" but "how do we decide *consistently* when decisions happen everywhere at once." Centralization and decentralization dodge that; structured coordination answers it.

When authority can't decide fast enough: the escrow problem

There's a failure that shows up in almost every large company scaling AI decisions, and it hides so well it deserves a name. I call it the escrow problem, because the decision gets stuck in a state where it's technically made but functionally unmade.

I watched it at a company running a lending model daily. The model analyzed applications and produced recommendations, with a sensible control on top: if its confidence fell below a threshold, a human had to approve before the decision was finalized. Except the model generated recommendations faster than humans could review them. By the time a reviewer looked, the context had moved (a customer's credit situation had shifted, market conditions had changed), and the recommendation that made sense when generated was now stale. The company faced two honest options: slow the model down, or let decisions move without full review. It did what most organizations do, which is neither. It raised the confidence threshold, expanded what the model could decide alone, added exception processes, added reviewers. It never solved the actual problem: the decision authority could not move as fast as the decision-generating process demanded.

That drift is expensive in three ways, all invisible until something breaks. It erodes decision quality: a human reviewing an hours-old recommendation is no longer asking "was this right?" but "is this *still* right?" — reactive, not deliberative. It creates drift with no decision behind it: what began as human oversight of all recommendations became something else through accumulated exceptions, with nobody ever choosing the change. And it creates liability: you *published* human review as your governance, so when a bypassed decision goes wrong your GC has to explain why your own process wasn't followed. "Our volume exceeded our capacity" does not survive a regulatory meeting.

Underneath, the escrow problem is a decision-rights problem: the organization never actually decided who holds authority. It sounds like it did ("the model recommends, humans review, we finalize"), but what it really said was "humans review *if there's time*, and then whatever happens happens." That's not authority; that's hoping someone with judgment is free. Organizations that handle this match authority to speed on purpose. *Fast* decisions (high volume, low stakes, proven track record): the model decides and humans monitor whether it's working rather than touching every case. *Medium-speed* decisions: the model recommends and a human approves or overrides on defined criteria, which only works if the human can keep pace, and when they can't, you're in escrow. *Slow* decisions (low volume, high stakes): humans decide with real time to think and the model just feeds them analysis. The trap is running medium-speed decisions through a slow process because you want the model's speed, the human's judgment, and full oversight at once. You can't have all three; pretending you can is how you promise three and deliver none.

So you choose what to give up, explicitly. And it's always one of three. Give up *speed*: run the model weekly instead of daily so review keeps up. Give up *human oversight*: let the model decide the high-confidence cases, backed by genuine monitoring and a real fallback. Or give up *some decisions*: don't deploy where you can't deliver fast human judgment, and keep the model as an internal tool rather than a decision system. Most organizations make none of these choices openly; they drift into giving up oversight while still claiming humans review everything. Don't say "humans review everything" when your volume makes that a fiction.

Governance upstream: where speed and safety stop being a trade-off

All of this eventually collides with the oldest fallacy in the governance conversation, that speed and safety are a trade-off: move fast and break things *with* governance risk, or move carefully and stay safe *but* slow. It's a false binary, and the best-run enterprises are disproving it. The trick isn't to accelerate governance or loosen standards. It's to shift governance earlier, so the decisions that would have slowed you down are the ones you never make.

Watch how the fallacy presents: "We need to move faster on AI. Governance is slowing us down. Can we relax some requirements?" Sometimes that's fair and your governance really is too slow. But often governance simply arrived too *late*: you already committed to an approach, built the infrastructure, promised the timeline, and now you're asking it to rubber-stamp a decision that's already locked. That's not a speed problem; it's a decision-architecture problem, and it hands governance an impossible choice: rubber-stamp (no safety) or block (no speed). There's no third option, because the deciding happened without governance in the room.

The organizations that move fast *and* stay safe put governance upstream, not as a gate but as a set of questions. Before a team settles on an approach, they workshop it with governance and risk experts present, and the question isn't "is this approved?" but "what breaks this? What are we assuming? What would make us regret this design?" That happens while architects are still choosing between approaches, before millions are spent, when changing direction is cheap. So the approval stage gets genuinely quick, because governance already blessed the *thinking*: no fatal flaw discovered at the gate, just an implementation to sign off.

The conventional architecture is a straight line (engineering builds, hits a governance gate, then deploys or gets blocked), with governance bolted on at the end, where the cost of change is highest. The architecture that produces speed with safety loops governance into the design instead, so the gate becomes a formality rather than a fight. That demands a different governance function (not one that shows up quarterly to audit, but embedded people who can reason about architecture, not just compliance) and engineers willing to invite them in early, which only happens when they trust governance is helping them build rather than hunting for things to stop.

This is why decision velocity is worth tracking. How long from "someone raised a governance question" to "we decided"? Three months means slow; three days means fast; three hours, because governance is embedded in the team, means you're leading: a structural signal that expertise is distributed and trusted rather than a bottleneck teams route around.

Because the true constraint on speed was never governance. It's decision quality. Deploy systems you don't understand and they fail in production; then you lose six months investigating and three more preventing a recurrence. That's not fast — that's fast *followed by* slow, and it costs more than spending a month getting the decision right the first time. The question was never "can we move faster without governance?" but "can we make better decisions faster by putting governance in the thinking?" Yes — but you bring them in early, not at the gate.

FIGURE 3.1

Route by Impact & Reversibility

Two questions set the route. Where a decision lands is the approval process.

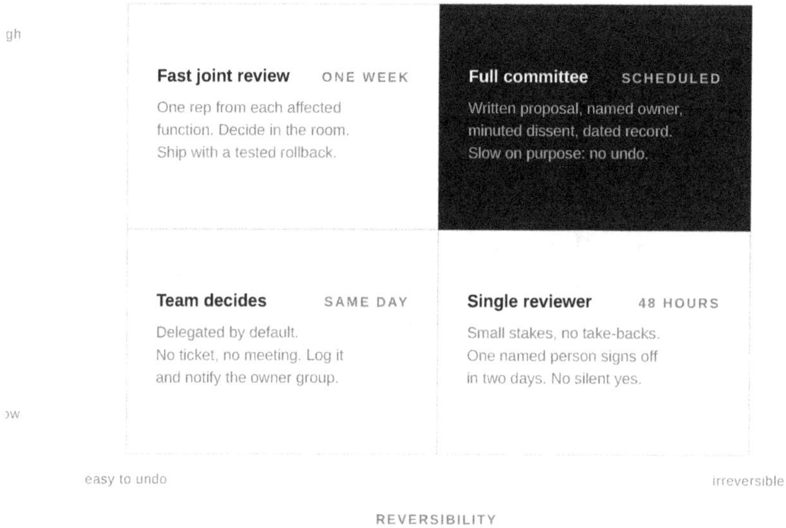

Irreversibility, not size, sets the route.

Small but unrecoverable still buys a reviewer. Large but reversible does not buy a committee.

Workshop use: teams write each proposed use case on a sticky and drop it in a quadrant.
Where it lands is the approval path — argue the placement, never the process.

FIGURE 3.1

FIGURE 3.2

The Decision Hierarchy

Three levels, three speeds — worked on a credit-limit model.

CONSTRAINTS FLOW DOWN ↓ | ESCALATIONS FLOW UP ↑

Enterprise — **Quarterly**, weeks to decide
BOARD AND EXECUTIVE COMMITTEE
Sets the outer limits: risk appetite, red lines, spend ceilings, what we will never automate.
SETS THE ENVELOPE
EXAMPLE May AI touch credit decisions at all? Under what cap?

Line of business — **Monthly**, days to decide
BUSINESS UNIT HEAD AND PRODUCT OWNER
Chooses which use cases run, and the trade-offs inside those limits — cost against accuracy.
MOST DECISIONS HERE
EXAMPLE Ship the credit-limit model to the SME book.

Team — **Daily**, hours to decide
PRODUCT, ENGINEERING AND DATA SCIENCE
Implements inside the envelope: thresholds, features, retrain cadence, rollout order.
WORKS INSIDE IT
EXAMPLE Raise the review threshold; retrain on last quarter.

Every decision belongs to exactly one level.

A decision that drifts up is one nobody at the top has time to make.
One that drifts down gets made without the constraints that bound it.

Workshop use: drop every pending AI decision onto the level that owns it.
Anything a level too high or too low is the finding — move it, or fix the constraint.

FIGURE 3.2

The Bottom Line

Neither centralizing nor decentralizing AI decisions works, because both answer the wrong question; the real work is placing each decision at the level that can make it fast enough and own its consequences. That means splitting authority (business units decide *what* to solve, a shared function sets *standards* rather than approvals, and audit *verifies*), then sorting decisions by altitude (enterprise, line-of-business, team) and by impact-and-reversibility. It means matching decision authority to decision speed on purpose, and moving governance upstream so the approval at the gate is fast because the thinking already happened. None of this is glamorous. It's standards, criteria, documentation, and the discipline to keep decisions at their right level instead of letting them drift up into a bottleneck or down into a shadow power structure. And there's a quick way to find out whether you've done it: name an AI decision your organization is currently arguing about, and listen to how the room answers who holds it and whether they can move at the tempo the use case runs at. If what comes back is "it depends," followed by a list of six functions, you don't have a governance framework — you have a coordination problem nobody has solved yet.

Field Kit

1. The decide-fast-enough check (run it after the Chapter 2 test, not instead of it). Chapter 2's test gets you a name. This one asks the thing that test doesn't: *can the named authority decide at the tempo this use case actually runs at?* Compare two calendars — how often the decision comes up, and how often whoever owns it is in a position to answer it. A committee meeting monthly cannot hold authority over teams shipping weekly; it can only hold the appearance of authority, while the real decisions get made by whoever needed an answer before the next meeting. When the tempos don't match, there are three honest moves: push the authority closer to the work, pre-authorize a band the team can act inside without asking, or slow the work to the speed of the authority. Named-but-too-slow fails as completely as unnamed, and it fails far less visibly.

2. The three-level decision sort (bring to your next governance design session). Put every recurring AI decision in a row and confirm it's made at the right altitude: - *Enterprise* (risk appetite, vendor strategy, data governance, budget, required processes) → slow, rare, cross-org leadership; should feel weighty. - *Line-of-business* (which initiatives, build-vs-buy, resourcing, refresh schedule) → fast, local, inside enterprise constraints. - *Team* (features, algorithms, training data, missing-value handling) → hours-to-days, no permission required. Any decision sitting too high is a bottleneck; any sitting too low is inconsistency waiting to surface.

3. The impact-and-reversibility grid. Classify each proposed model on two axes — *impact* (customer reach, financial exposure, compliance surface, integration dependencies) and *reversibility* (can we turn it off cheaply) — then assign the process: **low-impact/reversible** → team decides, notify only; **medium/partially reversible** → one technical + one compliance reviewer, parallel, 48-hour turnaround; **high-impact/irreversible** → full committee, capital-investment seriousness; **cross-boundary** → fast committee of only the stakeholders with skin in the game. Never route by size or "how strategic it sounds."

4. The escrow diagnostic (run before any human-in-the-loop deployment). Ask: "*What's the slowest decision process this use case can tolerate?*" Minutes → human review is NOT your control; monitoring and fallback are. Hours → staff enough human capacity to hold that pace. Days → a human-review process will work. Then check: is the model generating recommendations faster than reviewers can act? If yes, you're in escrow — choose explicitly to give up speed, oversight, or the use case. Drifting is not a choice; it's a liability.

5. The standard-vs-approval audit. Read your written AI standards line by line. Any rule that gates action ("must use an *approved* model," "must be *reviewed* by the committee first") is an approval in disguise — rewrite it as a verifiable statement of what good looks like ("must have a documented data-source rationale," "must have outcome monitoring," "must have a named accountable owner," "must have an out-of-distribution escalation process"). If you can't verify it after the fact, it's a gate, not a standard.

6. Decision velocity as a standing metric. Track time from "a governance question was raised" to "a decision was made," by risk tier: low-risk in days, medium-risk in weeks, high-risk longer only when analysis genuinely requires it. Three months = bottleneck, and teams are deciding around governance; three hours = governance is embedded and trusted. Review it quarterly at the enterprise level — rising velocity with stable incident rates means upstream involvement is working; falling velocity means governance has drifted back to the gate.

CHAPTER 4

Accountability Without Scapegoating

Part II · Decision Rights, Accountability & the Board

A financial services company shipped a credit decisioning model that started rejecting credit-worthy applicants from a specific zip code. When the incident finally got escalated, the conversation went like this:

"How did this happen?"

"The data was biased."

"Why was the data biased?"

"The historical training data reflected past lending patterns."

"Who owns the data?"

"The data team."

So the data team got blamed. The data scientist who built the model got blamed. The incident was closed as a "data quality issue," and everyone moved on.

But that is not what happened. What happened was a business decision — to train a credit model on historical lending data, knowing that data carried the biases of decades of past discrimination. Nobody made that decision consciously; nobody even knew it was a decision.

This is the accountability inversion, and you will find it in almost every company you walk into. We blame the people who execute (the data team, the engineers, whoever had their hands on the system last) and we excuse the people who decided what to build, which is to say everybody else. We do it systematically, and feel principled while we do it.

There is a better model, and it is not new. Airlines did not get safe by building systems to identify who made the error; they got safe by building systems to prevent errors from ever happening, and to learn from the ones that slipped through. That distinction is the whole chapter. Accountability that works is not a mechanism for locating a person to punish after the fact; it is a mechanism for making the system improve before it breaks, and then closing the loop so every failure teaches the organization something it keeps. Most enterprises say they want the aviation model, then build the opposite. This chapter is about why, and about the work of building the real thing.

Here is the reframe this chapter is built on.

The myth: accountability means tracing a failure back to whoever was responsible, so the right person answers for it.

The reality: accountability that works makes the system improve before it breaks, and closes the loop so every failure teaches the organization something it keeps.

The Last Hands on the Model

The failure was not in execution; it was in the decision about what to build. If the requirement was "build a model using historical lending data to predict credit-worthiness," and the data team said "this data contains systemic bias," and nobody changed the requirement, then the person who ignored that flag made a decision: to accept the risk of bias in exchange for a faster, cheaper model. A legitimate decision; you might even make it. But it is the business's decision, laundered into a technical failure the moment the model misbehaved. The trouble is that the people at the decision point do not experience it as a decision. They experience it as "running the process," while the data team experiences it as a constraint it flagged and got overruled on. So when something breaks, the convenient answer is always there: *the data was biased.*

This matters because companies optimize based on where they assign blame. Blame the data team and you will invest in better data science, never touching the real problem: a product, business, and risk decision to ship a model that discriminates against a protected class. Worse, you invert the incentives: the data team turns defensive and slow, and the actual decision-makers feel *less* pressure to think carefully, because the worst outcomes get caught downstream by the people who will take the fall.

Three People Made One Decision

Here is what should have happened, and it is almost boringly simple. Product says: "We want to predict credit-worthiness." Engineering and data come back with options, not an answer:

- **Option A:** Historical lending data. Fast and cheap, but it carries systemic bias from past discrimination.
- **Option B:** Transactional data and alternative credit indicators. Slower and more expensive, but less biased.
- **Option C:** Only recent lending data, accepting limited historical performance. Somewhere in between.

Then someone (product leadership, risk, or compliance) decides which option. *That* is the decision, the moment someone chooses which risks to accept: choose Option A and you are accepting discriminatory outcomes in exchange for speed and cost, a real choice, explicitly owned. What happens instead is that the choice gets bundled into "let's build the model." Nobody picks an option; the data team takes the fastest path and builds. When it breaks, you blame the execution, which never had the authority to make the call.

So who actually decided to accept the bias? Three people, none of whom felt like they were deciding anything. The **product person** who framed the problem as "predict credit-worthiness from historical data," and so chose Option A without noticing. The **engineering or data person** who translated the requirement into an approach and could have surfaced the trade-off. If they did not, they accepted it; if they did and were overruled, the decision moved up the chain. And the **business leader** who accepted the timeline and cost of the fast path. Nobody here is evil or incompetent. Three people each made a piece of a decision, none owned the whole thing, and when it broke everyone could honestly point at someone else: the data team blamed for being last in the chain, not for deciding anything.

Accountability worth having means two things: someone explicitly decided, and someone explicitly owns the outcome — genuine execution failures landing on the data team, predictably risky decisions landing on whoever made them, and nobody hiding behind "the data was biased."

The Blame Machine You Build by Accident

Now the harder problem: you can believe every word above and still build a machine that does the opposite. I watched a financial services company spend six months designing an AI governance framework whose centerpiece was an "approval log": comprehensive tracking of which stakeholder signed off on which model decision. They wanted accountability. What they built was a blaming machine.

When a model recommendation created a problem (a credit decision that looked discriminatory, a trading algorithm with unexpected volatility), leadership's first instinct was to trace the decision backward and find the failure point. Who signed off? Who did not ask the right question? The structure worked exactly as designed: it identified responsibility, and destroyed the incentive to ever flag a problem early. Documenting a concern before a decision created a permanent record of doubt; if something went wrong later, you would be asked why you raised it and did not escalate. Never write it down and you kept your deniability. The approval log became a liability trap.

People got good at *decision concealment*: diffusing a decision across enough people and processes that nobody could be held to it. "The model just flagged that." "We always do it that way." Which is worse than no accountability at all, because now nobody owns the decision or has the power to change it. The sharpest version I saw: a legal team flagged a genuine discrimination risk in a lending model, but the escalation process required the model owner to respond, and "I understand the concern and I'm choosing to proceed" would put their name on a paper trail. So the owner requested "additional validation": another review, more time, and the issue got absorbed into the standard refresh cycle, documented, then lost in process, then forgotten. The accountability structure made the organization *less* safe.

This is where aviation earns its place. Airlines do not convene accident review boards to blame pilots. They convene them because every incident, even a minor one, contains information about the system's weaknesses. Aviation calls this a blameless review, and the name is exact: the thing under investigation is the system that made the incident possible, not the human who operated inside it — because a system that punishes the person who reports goes blind. Organizations that handle AI accountability well ask structurally different questions:

- *What made this decision possible without someone catching it earlier?* (Not: Who should have caught it?)

- *What information wasn't available when the decision was made?* (Not: Who made a bad judgment call?)
- *What would have to be different for someone to feel comfortable raising this before it happened?* (Not: Why didn't anyone raise it?)

The difference is not rhetorical: a blame-oriented system makes the person who raises the concern a suspect; a system-oriented one makes them the resource that helped you avoid a bigger failure. The organizations that get this right run three overlapping mechanisms, explicit about which one answers which question. **Decision authority**: one person, not a committee, owns each decision, with the authority and information to make the call rather than rubber-stamp it; if nobody had authority to make it, the structure itself failed. **Escalation as information, not a stop sign**: raising a concern does not require agreement, it requires a documented response — someone with authority saying "I understand this and I'm proceeding" or "we're not," not "I'll think about it" or "let's form a subcommittee." **Post-decision learning**: when something breaks, the review feeds back into the governance structure itself. This is where nearly everyone fails: running the review, documenting findings, then changing nothing but a slide deck, so the same risk resurfaces six months later in a different costume. If the same risk appears twice, the governance structure failed; that is the diagnostic. The real choice is whether you want accountability that looks good in an audit meeting, or accountability that prevents the problem.

What You Wish You Had Written Down

There is one artifact that makes all of this real, and almost nobody builds it. When something goes wrong, someone from compliance or the General Counsel's office asks: "Why was this decision made?" And usually nobody knows. There is no record: scattered emails, an archived Slack channel, and the person who made the call left six months ago. Now you are arguing retroactively about whether the decision was reasonable, and nobody can reconstruct the reasoning.

The artifact is the decision record from Chapter 1, filled in. Organizations build decision-making *processes* (governance boards, approval workflows, sign-off sheets) and mistake the process for the record. Here is the difference, for a model deployment:

> ***Decision:*** *Deploy the demand forecast model to production for the northeast region, phased.* ***Decision-maker:*** *VP of Operations. (Input from the Chief Data Officer and the Finance Director; the call is hers.)* ***Date:*** *November 2024* ***Why:*** *Improves forecast accuracy by 12% versus the manual process, cutting inventory carrying costs. Projected ROI of $2.3M annually, payback in 8 months.* ***What we are accepting:*** *Accuracy is good but not perfect: a bad forecast means the wrong inventory in the warehouse. We have only 18 months of data and have not seen the model run through a full seasonal cycle.* ***Mitigation:*** *One region first, so a rollback costs us a quarter and not a year. Manual review on every order above $1M for the first 90 days.* ***Recalibrate / stop:*** *Weekly forecast accuracy below 85% two weeks running, recalibrate. Below 80%, or any stockout traceable to the model, stop and revert to manual.* ***Monitors:*** *Demand planning lead, forecast accuracy, weekly. Finance Director, carrying-cost variance, monthly.* ***Escalation:*** *Either monitor goes straight to the VP of Operations, who answers within five business days.* ***Review:*** *90 days (go/no-go), 6 months (full rollout).*

Now run the credit model from the opening of this chapter through the same form and watch the accountability inversion become impossible to stage. *Decision: train the credit model on historical lending data. What we are accepting: the model will reproduce the lending patterns of the last twenty years, including the discriminatory ones. Monitors: approval-rate parity by zip code, monthly. Decision-maker: VP of Product.* Nobody writes that sentence and ships it unchanged. That is the whole mechanism: the record does not stop you from choosing Option A, it stops you from choosing Option A without noticing, and it puts a name on the choice before the failure rather than after, when the only name available is whoever touched the model last.

If something goes wrong, the story is then not "we made a reckless decision" but "we made a decision with these safeguards and the safeguards did not work as expected." The second story is defensible; the first is not. Most sign-off sheets capture the decision and the rationale and stop there, skipping the fields that carry all the weight under scrutiny (what you were accepting, and who was watching for it), so you end up thinking you were monitoring something you were not, or that you had a contingency that depended on someone who no longer works here. You need this not for "which hyperparameter should we test," but for the consequential calls (deploy to production, raise the confidence threshold, change the fairness criteria), written while the decision is fresh, not retroactively when you are constructing a story and are probably wrong about half of it.

When the System Makes the Mistake

Sometimes the failure is genuinely one person's: someone did not monitor the model, skipped the edge case, wrote a rule too broad. One clear failure, one fix; name it and move on. But usually it is not that simple, and this is where scapegoating does its real damage. A model starts misbehaving, and when you trace it back you find: it was trained on data collected in a way nobody realized would matter; the monitoring was built to flag one kind of drift and it drifted the other way; the testing covered normal conditions, not the edge case that broke it; the deployment had no kill switch for this failure mode. Three people built the model, two validated it, two deployed it, the data team prepared the inputs, ops monitored it, product interpreted the output. Seven people. No single one made the mistake. The system made the mistake.

But accountability frameworks are built for "this person failed to do their job," not "the system made a mistake." So you pick someone, reprimand or retrain or reorganize, and tell yourself you restored accountability. You did not. You taught everyone to hide problems earlier next time, pass the hot potato, or document their own work so exhaustively that blame cannot land on them individually. You made accountability mean "not getting caught."

Real accountability here runs in four moves. **First, separate investigation from punishment**: trace back through the system, find where it made assumptions that turned out wrong and what information was not available, and distinguish whether someone actively broke a rule or the system made it too easy to break one by accident. The goal is to understand, not to punish. **Second, identify the system failure, not the person failure**: in the case above, any one of those fixes (validating the data-collection condition, catching the actual drift, testing the edge case, a kill switch) would have prevented the incident. **Third, make the owner of the system responsible for fixing it**: not "you failed," but "the system you own failed, here is why, and you own the change," because you have the domain expertise and the authority. That is accountability without blame. **Fourth, create a forcing function**: the step almost everyone skips, because once you name the fix and the owner, the pressure lifts and it slides. It needs a due date, a check-in, and an explicit sufficiency decision: "The data team owns the fragile data-collection process. You have 30 days to design a fix; on day 30, walk us through it and we approve or iterate." If it is not built by day 60, it escalates to leadership, not as blame, but as "this fix is not happening and we need to solve it together." The system stays broken only if someone actively decides to accept a broken system.

The dividend is trust. If accountability means blame, people work to avoid it: defensive, risk-averse, over-documenting, escalating so someone else takes the fall. If it means "your system failed, help us understand and fix it," they surface problems fast and help design the fix. That is why this framework survives a crisis: you can only learn fast if people feel safe enough to tell you the truth.

Closing the Loop

Most organizations stop at accountability. Someone is responsible for a decision, and that is the end of it. But accountability without a feedback loop is just punishment: it assigns responsibility, never improves the system, and teaches the next person that the safe move is to avoid deciding at all. Accountability says: you made a decision, you are responsible, if it was bad you will face consequences. A feedback loop says: you made a decision, we measured the outcome, here is what we learned, here is what we will do differently. The first defers the problem; the second makes the system smarter — because it fixes the process that allowed the failure, not the person who happened to be standing over it.

Building the loop takes four things, none of them glamorous: **measure the outcome** (did retraining actually improve accuracy, by how much?), **reflect** (why did this happen, were there unintended effects?), **incorporate the learning** into the next decision, and **document the loop** as knowledge, not blame. On a real model: accuracy drifts, you retrain, it improves from 89% to 92%, and most teams stop there. The loop keeps going. Why did it drift at all? You measure the data distribution and find that demographic group A, 40% of the training data, is now 25% of the live data. That shift dragged down accuracy for some groups even while the aggregate looked stable. The lesson (when distributions shift, aggregate accuracy goes blind to real drift) sends you to track demographic-specific accuracy for every model and update the governance standard. Next time, you catch the drift earlier because you are finally monitoring the thing that actually moves.

Nobody does this by default, for human reasons: it is not mandatory, it feels like overhead, it makes people defensive about reopening a bad call, and the payoff is slow. None of that changes the fact that organizations which skip it are making decisions the same way in year three as in year one. So start small. Pick one decision type (retraining is ideal), document why, measure the result, reflect, and write down the lesson. After three to five cycles you will have patterns you can turn into sharper criteria, then expand to deployment and prioritization decisions. Feedback loops also let you fire your own bad rules: mandate fairness audits on every model, measure a year later what they actually caught, and if the answer is "very little, mostly bureaucracy," change the rule (high-stakes models only), because the process is now built to evolve.

Who Owns the Learning

One structural choice determines whether any of this compounds: where accountability lives. When it sits with a central governance team that owns risk but cannot change how products get built or what data gets used, accountability floats free of action; the team checks boxes, the business moves fast, and nobody really learns. That is compliance theater, and it does not compound: each team decides independently, learnings do not spread, and the organization makes the same mistake at three different scales. Push accountability out to the teams that can actually change something and you also get the only conditions in which blame-free accountability is possible, because a team investigating its own system has nothing to gain from a scapegoat: the failure landed on them, the fix is theirs, and an honest diagnosis is self-interest rather than exposure. Chapter 16 covers how you build that capability across an organization; the point here is narrower and worth a crude measure: count how many governance calls in the last quarter were made by the team that owns the system versus escalated to a central function, and if almost everything escalated, your accountability lives somewhere that cannot fix anything.

That kind of ownership needs a few structures to hold. Every significant system has a real, substantive owner who understands the assumptions it rests on and the conditions that would make it unsafe, documented at deployment time along with when they will be tested. When something fails, the investigation starts by asking what the owner knew and when, not to blame, but because the answer routes the learning: a risk consciously accepted taught you your risk tolerance; a risk identified and then missed means you fix your risk-identification process; a genuinely unforeseeable risk is organizational education. Run that as a conversation and not a checklist: checklists in this territory decay within two quarters, because the useful version of every question is the follow-up, and a form cannot ask a follow-up. Above the individual incident, connect the patterns: if three teams make similar risk decisions three different ways, or the same class of problem surfaces in three systems, that is a governance signal. The organizations that pull ahead aggregate these patterns and make them visible — a quarterly review, or a dashboard showing the risks discovered, the systems affected, and whether the learnings are applied.

By year two of an enterprise AI program, you can usually tell which model an organization is running. The learning-oriented ones have systems that are stable and improving, with fewer incidents over time; the theater-oriented ones have constant fires and constant surprises, rediscovering the same problems in new forms. The difference is not how rigid the governance is. It is whether the learning loops actually close.

FIGURE 4.1

The Accountability Inversion

One incident, read two ways — the same credit-limit model, six weeks wrong.

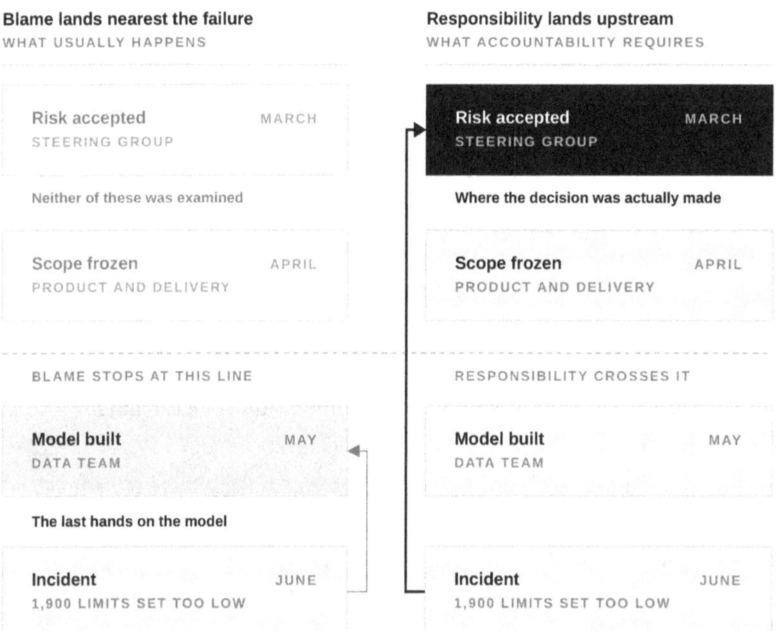

Blame travels one hop. Responsibility travels the whole chain.

The data team built exactly what was approved. Firing them changes nothing — the same risk is still accepted upstream, by the same people, next quarter.

Workshop use: map one real past incident onto both panels — where blame landed, and where the decision actually lived. The gap between the two arrows is the work.

FIGURE 4.1

The Bottom Line

Accountability fails when you treat it as a search for the person to blame, and works when you treat it as a mechanism for the system to learn. The organizations that get AI governance right borrow aviation's instinct: build systems that prevent errors and surface them safely, not systems that trace failures backward to a name. That means separating the decision from the execution so the right person owns the call, capturing the reasoning while it is fresh, investigating failures as system failures with a real forcing function behind the fix, and closing the loop so every incident sharpens the next decision. Filled-in decision records, documented escalations, due dates, quarterly pattern reviews: none of this is glamorous, which is exactly why most organizations skip it and keep blaming the data team. Do the boring work, distribute the ownership, and your governance compounds instead of repeating itself.

Field Kit

1. The incident-review question set. Bring these to the room instead of "who messed up": - What made this decision possible without someone catching it earlier? - What information wasn't available when the decision was made? - What would have had to be different for someone to feel comfortable raising this before it happened? - Walk me through the assumptions at each stage — where was the information incomplete, and where were the processes fragile?

2. The decision → execution split. Before any consequential model ships, require the team to present options with trade-offs (fast/cheap/biased vs. slow/expensive/cleaner), name the one person with authority to pick, and record which risk they are accepting. If nobody has authority to make the call, the structure has already failed — fix that first.

3. The decision record, filled in while it's fresh. For any consequential decision (deploy to production, raise a confidence threshold, change fairness criteria), fill in Chapter 1's record at the moment of the call, not afterward. Then audit it on one narrow test: if it names the decision and the rationale but not what you were accepting and who is watching for it, you don't have a record — you have a sign-off sheet, and after an incident it will tell you only that somebody approved something.

4. The escalation-response rule. A raised concern doesn't require agreement; it requires a documented response from someone with authority — "I understand this and we're proceeding" or "we're not." Ban "I'll think about it" and "let's form a subcommittee" as closures. An escalation raised and dismissed is a decision; one never logged is a liability.

5. The forcing-function template. Every system fix gets four things or it doesn't count: an owner with authority to fix it, a due date, a check-in, and an explicit sufficiency decision. If it's not built by the second checkpoint, it escalates to leadership — as a shared problem to solve, not a blame to assign.

6. The learning-loop starter. Pick one decision type (retraining is ideal). For each instance, document the trigger, measure the outcome, reflect on why it worked or didn't, and write down the lesson. After three to five cycles you'll have patterns you can turn into criteria — and proof the loop is worth extending to deployment, audit, and prioritization decisions.

CHAPTER 5

The Board's Job: Oversight, Dashboards & Risk Appetite

Part II · Decision Rights, Accountability & the Board

Accuracy: 95.2%, up from 94.8% last quarter. False positive rate: 12%, down from 14%. Inference time: 42 milliseconds, steady. Monthly predictions: 1.2 million, up 18%.

That is the quarterly AI slide at a regional health insurer, and it is the slide at most of the boards I have sat in front of. Every number on it is true. Not one of them tells a director whether to approve anything.

At that insurer, a director at the far end of the table asked the only question that mattered. "So what could go wrong?" Nobody answered, because nobody actually knew.

That silence is the subject of this chapter. The slide was not dishonest and the AI lead who put it up was not incompetent. He was showing the board the dashboard his team used to *run* the model, which is a perfectly good dashboard for running a model and a useless one for exercising fiduciary duty over it. A board is not a second engineering review. It is the body that has to say, in advance, which failures this organization is willing to live with, and then know whether it is living with them. (Chapter 6 takes up the parallel failure inside the validation process itself; this chapter stays in the boardroom.) So: the decision-grade information a board actually needs, the risk appetite it should actually be signing, and the governance that has to survive the moment someone powerful doesn't like the answer.

Here is the reframe this chapter is built on.

The myth: A board oversees AI by reviewing the model's performance numbers, so a dashboard of accuracy, error rates, and volume is what fiduciary duty looks like.

The reality: A board is not a second engineering review; its job is to decide in advance which failures the organization will live with, then know whether it is

living with them.

The Dashboard That Answers the Wrong Question

Six things belong in front of a fiduciary body. Accuracy, latency, and volume are none of them.

Start with the cost of being wrong, because it is the number the accuracy rate is hiding. A board doesn't need the error rate; it needs both directions of error priced in dollars, and the name of the person who chose which one the model is currently tuned to avoid. Ask a governance committee what a single wrong call costs the company and watch what happens. At the insurer I got four answers from four executives and not one of them had a figure attached. That 12% had been reported to the board every quarter for two years, and nobody had ever converted it into money. A board cannot oversee a rate it cannot price. Chapter 6 covers how those two dollar figures get built and why they are almost never equal; the board's job is to refuse the slide until they exist.

Then ask what has changed about the world the model runs in. This is distribution shift, and it almost never makes it onto a governance dashboard even though it belongs at the top. A model trained on pre-pandemic customer behavior can still be operating perfectly years later — on assumptions that no longer hold. Nothing on the accuracy slide will tell you the ground moved. The board needs to know the inputs are drifting, or the outcomes are changing, or the relationship between the two is degrading, *before* you spend another quarter optimizing for patterns that don't exist anymore.

Third: where is the model breaking? A blended accuracy number is not a governance number. If the model runs at 97% for one customer segment and 72% for another, the board needs to see both, stratified by the segments that actually matter to the business. The aggregate is the number that lets everyone in the room feel fine.

Fourth: if it went wrong, how fast would you know? Call it mean time to detect, and make someone own it as a reported figure rather than an aspiration. A degradation you catch in two days is an incident you manage. The same degradation sitting undetected for three weeks is the difference between a problem that touched thousands of customers and one that touched millions. Detection lead time is a governance metric, and almost nobody reports it to a board.

Fifth: can you explain a decision? Not in weights and feature importances. In plain terms: what the model saw, how it weighted the information, what it decided, and whether we think that was right. If a regulator or a customer asks, and only 85% of your decisions are explainable within your own standard, you have a named, sizeable risk exposure, and the board should see that percentage.

Sixth, and most overlooked: how much are you actually trusting the model? What share of decisions is it making alone, and how often are humans downstream overriding it? If humans override the model 40% of the time, you either have a trust problem or a model problem. Either way you are not getting what you paid for. This is the metric that separates *adoption* from *actual adoption*, and it never appears on the technical dashboard because it isn't a technical fact. It's a governance fact.

None of these are glamorous. They require logging which decisions turned out right and which wrong, tracking demographic breakdowns of performance, measuring the gap between degradation and detection, documenting override rates: operational hygiene that no one demos with pride, which is why organizations skip it. The ones that moved from compliance reporting to genuine risk management all did the same unremarkable thing: they stopped treating AI metrics as a data science problem and started treating them as a business accountability problem. Your accuracy metric tells the board about your model; these six tell the board about its exposure. If you can only put one set in front of them, it's the second.

What Your Board Is Actually Signing

A dashboard tells the board what is happening. A risk appetite statement tells them what they agreed to tolerate before it happened. And here most boards have a document that is worse than useless, because it creates the impression of a decision without the substance.

You have seen the language. "We maintain a balanced approach to risk while pursuing innovation." "We accept calculated risks that support strategic objectives." Drafted by the Chief Risk Officer's team, worded so that any outcome can be called compliant. Then the problem arrives (a model fails, a bias issue goes public, a regulator gets serious), and nobody can point to that statement and say, "Yes, this is exactly what we authorized." That isn't a risk appetite statement. It's governance theater, an aspiration wearing the costume of a constraint, describing what the organization hopes to achieve rather than what it is willing to tolerate.

A real one names, in operational language, which failures the organization is willing to experience and which it isn't. Not failure in general: specific failure modes, specific thresholds, specific trade-offs. On first read it should feel uncomfortably concrete.

Because the thing you're actually deciding is sharper than most boards realize. AI will produce outcomes a rule-based system wouldn't: accurate on average and wrong for specific people, failing in ways that are hard to predict, sometimes doing worse than the legacy alternative it replaced. So the question underneath the appetite statement is: *under what conditions will we accept outcomes we wouldn't accept from a traditional system?* That is not a philosophical question. It's an operational one.

Take hiring. A traditional rubric (minimum GPA, certain schools, years of experience) is crude and misses talented people, but it's consistent, and when it fails you know why. An AI hiring system can be more accurate across the whole population while being opaque for any individual candidate, and it will sometimes make a mistake a human would have caught. What is your appetite for that trade? Not in principle — in practice. Do you accept a model that's 5% more effective on average but makes occasional rejections you can't explain to a candidate? Ten percent? Twenty? At what accuracy threshold do you switch back to rules? Your General Counsel already knows these are the questions that will matter when the litigation starts.

A statement worth something reads like this: "*We accept hiring models that outperform our traditional rubric by at least 8% on overall placement success. We do not accept opaque rejections for protected-class candidates. We conduct quarterly bias audits and halt model use if disparate impact on any protected class exceeds 3 percentage points.*" Specific, defensible, and something a board is genuinely committing to, rather than a sentence that fits any future.

The other mistake boards make is writing one appetite and applying it to everything. An acceptable failure for a recommendation system (a wrong suggestion the user ignores) is not acceptable for a compliance system, where a false positive detonates audit cost. Separate your failure modes by consequence, because each carries a different appetite:

- **Direct customer or citizen impact.** Loan decisions, hiring, benefits eligibility, medical triage. Failure mode: accuracy and bias. The question needs a numerical answer: what accuracy is required before you deploy, and what population disparities you'll tolerate.

- **Operational efficiency.** Contract review, log analysis, pattern detection. Failure mode: missed problems and rework. The question is the actual error rate that flips you back to human review, not "we'll monitor it."
- **Strategic or regulatory exposure.** Compliance, audit, regulatory reporting. Failure mode: misclassification with enforcement consequences. Do you accept any error rate at all, or require 99.9% accuracy with human review of every flagged item, and regulatory pre-approval before deploy?
- **Speed and convenience.** Chatbots instead of agents, automated triage instead of manual. Failure mode: a bad experience or an escalation. What escalation rate is acceptable, and how much productivity gain justifies how much friction?

Setting a real appetite across these categories changes three things. It forces alignment *before* deployment, so when someone later asks "should we keep using this despite the bias issue," there's a reference point instead of a fight. It clarifies the monitoring you need: if the appetite says 5% error is acceptable, you build monitoring to catch 4.5%, not perfection. And it makes decisions delegatable: once the board sets the appetite, operations and compliance can evaluate models against it without escalating every call. That is the move from review-based governance to criteria-based governance.

The hard part was never writing the statement. It's defending the numbers. Why 5% and not 3%? Why 99.9%? The insurer's appetite statement, when I finally got a copy of it, ran two sentences and contained no numbers at all. What eventually fixed it was not a redraft; it was an argument. Their Chief Medical Officer said he would not accept a triage model that wrongly cleared more than one case in two hundred. The COO said that at that threshold the model saved nobody any time and they should switch it off and hire more nurses. They spent most of a meeting on it and came out at one in a hundred and fifty, plus a weekly manual read of every case the model cleared at low confidence. That number is worth something precisely because two executives who each had standing to be right fought over it in front of the board. A statement that arrived through polite compromise means nothing when you need it. And if the board won't commit to specific numbers at all, that is information too: the organization isn't ready to deploy the model, and that is a decision far cheaper to make now than after something breaks.

When the Rules Meet Power

So you've built the good structure. Clear decision rights, real documentation, regular review cycles, smart people in the room. It works for about six months. Then a board member asks a question that matters ("Why did we turn off that model?" or "Why didn't we deploy this system?"), and the honest answer doesn't line up with what someone powerful wanted. Now your governance structure stops being an asset and becomes a problem.

This is where governance gets political. Not backroom dealing, though that happens — political in the plain sense of power. Governance is a set of rules, rules constrain power, and when a rule blocks someone with actual authority, it either gets changed or gets ignored. Most organizations design governance for a world where the structure is neutral and everyone respects it. That world does not exist.

The failure shows up in one of four shapes. **The override:** someone with authority needs something, the structure says "not yet, wait for approval," and they proceed anyway; governance becomes advisory. **The reinterpretation:** the structure says "no deployment without a fairness audit," and someone argues "fairness audit" is ambiguous (what counts, who decides), and their model is fair by their definition; governance becomes interpretable. **The work-around:** the structure applies to production models but not to pilots, so everything becomes a pilot; governance becomes inapplicable. **The evolution:** the structure works, the business changes, it's now "too slow" and gets relaxed, then something breaks and it gets tightened; governance becomes reactive instead of stable. In none of these cases did the structure fail because it was badly designed. It failed because it wasn't designed to survive pressure.

The structures that do survive share three properties, and none of them are about writing better policy.

First, they align incentives. If following the rule creates friction without benefit, people route around it; if it creates benefit, they defend it. The strongest governance is the kind where obeying the rule produces an obviously good outcome for the person obeying it: "We do fairness audits before deployment. This prevents lawsuits. This saves us money." A structure that produces compliance benefit but no business benefit for the people it constrains is vulnerable, no matter how well it reads.

Second, they create accountability that actually bites. In most organizations, following the structure is invisible and ignoring it carries no immediate consequence — a structure engineered to fail. Durable governance gives the people inside the process skin in the game: if a model ships without proper review and causes harm, everyone knows who approved it and who was overridden. The risk is shared, which is exactly what makes someone think twice before the next override.

Third, they match governance weight to reversibility. Some decisions can be undone: deploy a model, watch it, roll it back. Some are hard to undo: pick a vendor, build the integration, hire for the skill. Some are nearly impossible: train a model on proprietary data, set a regulatory precedent. Governance can be light for the reversible and heavy for the irreversible. Make everything heavy and you're not making trade-offs, just slowing everything down until the structure gets resented and bypassed; make everything light and nothing stops a bad irreversible call. Proportionality is what keeps people from experiencing the whole apparatus as an obstacle to be routed around.

One diagnostic question reveals which kind of governance you actually have: *if someone with authority disagrees with what this structure says, what happens?* If the honest answer is "they override it," your governance is advisory, which might be fine, but stop pretending you have authority you don't. If the answer is "we discuss it and follow the structure," ask why: usually because the people with power understand why the rule exists, see benefit in it, or carry accountability if it's ignored. Governance that says "we have policies and you have to follow them" eventually loses to someone who doesn't want to. Governance that says "here is why this structure protects you and creates value for you" is the kind that lasts.

The Twelve-Minute Conversation

All of this (the dashboard, the appetite, the durable structure) depends on a board that actually engages. I have heard the following exchange, nearly verbatim, in three different boardrooms.

"Do we have an AI governance framework?" Yes. "Do we have oversight?" Yes. "Are we compliant with regulations?" As far as we know. "Any board members have concerns?" No. "Great, let's move on."

Twelve minutes. That is not a board conversation about AI governance; it's a checkbox conversation. Boards ask these questions because they're the ones they know to ask, imported from data governance and cybersecurity and compliance, where they're good questions. For AI they are nowhere near sufficient, and the real conversation is longer, more uncomfortable, and almost nobody is having it.

Here is what a rigorous version sounds like.

"Walk me through a model that went wrong. How did we catch it, what did we do, and what did we learn?" This question reveals everything. The right answer is concrete: "A model was drifting on customer segment X. Our drift monitor flagged it inside two days, we routed that segment back to human review the same week, retrained, and here is the segment-level alert we added so the next one gets caught before it reaches production." The wrong answers are "we haven't had a model go wrong yet" or "it would take a while to trace back." Note that an *ugly* number, honestly reported ("we found out three weeks in, from a customer complaint"), is a better answer than either, because it is at least a real incident with a real detection lag attached to it. If the board has never walked through a real example, they don't understand the organization's risk posture; they've only been told it's fine.

"How do you know that all the models in production are actually the models you think are in production?" A question about operational reality. You supposedly have a hundred models deployed. Do you know what half of them do, who owns them, whether they're still running or were shut off months ago? Many organizations have shadow models serving live traffic nobody is officially monitoring. The board should know whether yours is one of them.

"When a model makes a decision that looks wrong, who investigates, and how long does it take?" This is operational readiness, and it's where your detection number either survives contact or doesn't. Flagging drift in two days buys you nothing if the investigation queue is four weeks deep and the model keeps deciding the whole time it sits there. The answer a board should want names a person, a clock, and (the part almost everyone omits) the authority to pull the model out of the loop while the investigation runs.

"Tell me about the largest financial exposure if a model failed silently. Walk me through the scenario." Risk quantification, and your CFO's question as much as anyone's. You should be able to say: if this model failed silently and nobody caught it for three weeks, the impact would be $50 million. Or $5 million. Or $500K. But you

should know the number. If you don't, you don't understand your risk, and if you don't understand it, you aren't managing it. The same arithmetic runs on the spending side (Chapter 15 takes up what the board is getting for its AI budget), but note here which line is reliably the starved one: monitoring, because nobody sees why it matters until it is the thing that would have saved you. Its harder cousin: *"Walk me through how you'd shut this organization's entire AI operation down in 60 days, and tell me what would be hard."* Most teams have never thought it through — and discover their AI operation is fragile in ways they never mapped.

"Show me the risk appetite we're operating under, and tell me which number in it somebody argued about." This is the board auditing its own work, and it is the fastest way to find out whether the document from earlier in this chapter is a constraint or a costume. If the statement contains no numbers, the board never made a decision; it approved a mood. If it contains numbers but nobody in the room can remember the argument that produced them, someone drafted them to be approvable rather than to be true — and they will not hold the first time a powerful person wants them not to.

Why don't boards ask these? Four reasons, none of them stupid. Comfortable questions are easier than "do you actually understand your largest risk." AI expertise is genuinely rare on boards: they understand cybersecurity because they've lived with it for a decade, so they default to "do we have a framework." The executives would rather not be asked: if the CIO knows the answers are thin, the last thing they want is the board probing. And boards haven't calibrated what good looks like, so "we have a governance committee" gets accepted as sufficient because nobody in the room knows whether it is.

When boards start asking the harder version, the organization changes underneath them. Problems become visible faster, because the questions force executives to name what they don't know. Resources move toward the parts everyone was underfunding: monitoring, ops. And accountability becomes real: when the CIO commits to the board that "we have a process to catch model failures within 48 hours," they go build that process. Without the commitment, it stays a slide.

That is the whole reason any of the boring work gets funded. Executives invest in operational maturity when the board asks about it, and not reliably before: board pressure is what makes someone build the 48-hour detection process, stratify the accuracy number, and put a dollar figure on the worst case. Which is also the test, and it runs the opposite way to instinct: a smooth meeting is the bad outcome. When the board conversation is comfortable, you're not doing oversight. You're performing it.

FIGURE 5.1

The Portfolio Risk Quadrant

Every production system gets a dot. Only one quadrant gets your attention this quarter.

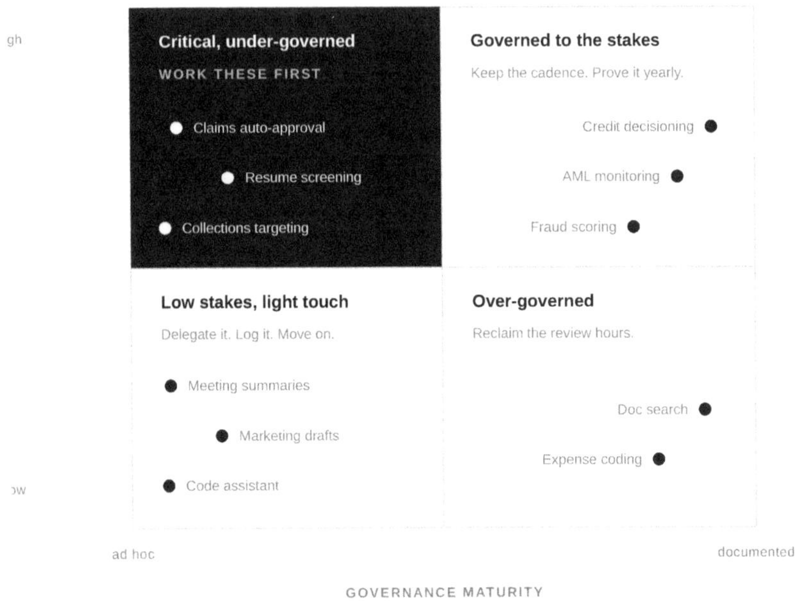

Governance follows the regulator, not the stakes.

The audited systems are governed. The ones that quietly decide about people — who gets hired, paid, chased, denied — are the ones nobody owns. That gap is the whole exercise.

Workshop use: teams dot their live portfolio, then defend every placement out loud. Every dot in the dark quadrant leaves the room with an owner, a bound, and a date.

FIGURE 5.1

The Bottom Line

Boards are routinely handed a dashboard built for a data science team (accuracy up, latency down, volume climbing) and asked to exercise fiduciary oversight with it, which is why nobody can answer the moment someone asks what could go wrong. Real oversight runs on three things the vanity dashboard can't provide: decision-grade risk metrics (cost of being wrong, distribution shift, stratified performance, time-to-detect, explainability, override rate), a risk appetite statement made of specific failure modes and contested numbers rather than balanced-approach boilerplate, and governance designed to survive the moment someone powerful dislikes the answer (aligned incentives, accountability that bites, weight matched to reversibility). All of it is boring operational hygiene, which is exactly why organizations skip it. The board is the forcing function that makes the boring work happen; without a conversation that takes an hour and makes someone uncomfortable, governance stays theater. Do the boring work, and you get oversight that holds when it's tested instead of a checkbox that gets checked in twelve minutes.

Field Kit

1. The board risk dashboard — six metrics that replace the vanity slide. For every model touching a consequential decision, report: (1) cost of being wrong, priced in dollars for each direction of error (Chapter 6's error-cost worksheet is how those two figures get built); (2) distribution shift — are inputs, outcomes, or their relationship drifting; (3) performance stratified by the segments that matter, never the blended number; (4) mean time to detect degradation; (5) percentage of decisions explainable within your own standard; (6) human override rate. If the deck only shows accuracy, latency, and volume, it was built for the wrong audience.

2. The four-category risk appetite template. Write a separate appetite statement for each failure class — direct customer/citizen impact, operational efficiency, strategic/regulatory exposure, speed/convenience — and force a number into each: required accuracy, tolerated disparity, the error rate that triggers a switch back to human review, acceptable escalation rate. One appetite applied to all four is the tell that no real decision was made.

3. The number-defense drill. For every threshold in the appetite statement, answer "why this number and not a tighter one?" aloud. A number that survived real disagreement — one executive wanted one wrongly cleared case in two hundred, another said that threshold makes the model pointless, you landed at one in a hundred and fifty with a weekly manual read — is a constraint; a number nobody contested isn't one yet. If the board won't commit to any number, treat that as a signal the model isn't ready to deploy.

4. The governance-durability diagnostic. For each rule, ask: "If someone with authority disagrees with what this says, what happens?" Then three follow-ups — Does following it create benefit for the powerful people it constrains, or only friction? Does ignoring it carry a visible, shared consequence? Is the weight matched to how reversible the decision is? Any rule that only produces friction, carries no consequence, or over-governs a reversible call will not survive its first political test.

5. The questions to bring to the board meeting. Put these on the agenda and budget an hour: Walk me through a model that went wrong — how did we catch it, how long did it take, and what did we learn? How do we know the models in production are the ones we think are? When a model makes a decision that looks wrong, who investigates, how long does it take, and who can pull it out of the loop meanwhile? What's our largest silent-failure financial exposure, in dollars? Show me the risk appetite we're operating under, and tell me which number in it somebody argued about. Walk me through shutting the whole AI operation down in 60 days — what's hard? If every answer is comfortable, you asked the wrong questions.

6. The "model that went wrong" readiness check. Before the board meeting, have every model owner draft the concrete answer to question one: the incident, the detection lag in days, the fix, and the monitoring change it produced. A team that can't produce a real example, or says "it would take a while to trace back," has just told you the risk posture is unknown — which is the finding, not the failure.

PART III

Model & Data Risk

CHAPTER 6

When Validation Measures the Wrong Thing

Part III · Model & Data Risk

It was a Wednesday afternoon in late October when someone finally noticed.

A risk officer was working through a batch of decisions from a lending recommendation system. The system had been live for eight months. It worked. The models had been validated, backtested, tested on holdout data. The metrics looked good — accuracy stable, validation healthy, monitoring clean. But in this particular batch, something didn't add up. There was a pattern of approvals that shouldn't have been approvals.

The system hadn't broken. It wasn't defrauding anyone. It had simply learned to optimize for the thing it was rewarded for — approval velocity — and in doing so it had learned to weight certain borrower profiles in a way that made internal sense and created external risk. When they dug in, they found that over eight months the system had recommended approval on credits whose characteristics, under scrutiny, implied a higher default probability than historical norms. The organization had been quietly taking on tail risk that nobody had measured, because nobody was measuring the right thing.

The cost, when they finally calculated it, ran into the millions. Not because the model was wrong. Because the model was right about the wrong thing, and every gate it passed through was designed to confirm that it was right.

This is not a story about one bank. It's a pattern that maps across domains, risk types, and organization sizes, and it turns on a distinction most enterprise leaders chronically underestimate. Ask a room how they know a model is good enough to deploy, and they answer with metrics: accuracy, F1, AUC, precision and recall. So they build validation processes optimized for those numbers, watch them carefully after deployment, and feel validated when they hold. Here is the problem in one line. You're measuring whether the model *works*. You should be measuring whether the decision it produces is *acceptable*. Accuracy tells you the model works. It does not tell you that deploying it is a good business decision, and the gap between those two questions is exactly where the multimillion-dollar failures hide.

Here is the reframe this chapter is built on.

The myth: if a model's accuracy and fairness metrics look good, validation is done and the model is safe to deploy.

The reality: validation has to test whether the decisions the model produces are ones your business is willing to make and defend, and no metric can tell you that.

The Accuracy Trap

Start with accuracy, because it's the easiest one to get wrong.

You've trained a model to predict something. It's 95% accurate on test data, 94% on a fresh holdout set, 93% after the first month in production. These are good numbers. And they are, on their own, almost meaningless for understanding your risk.

What matters is what happens when the model is wrong. If it's predicting churn and it's 93% accurate, it's wrong 7% of the time, and that 7% splits into two very different piles: people it flagged as churn risks who stayed (false positives), and people it cleared who left (false negatives). These costs are not symmetric. A false positive means you treat a loyal customer as a flight risk: you send them a discount, some outreach, special handling. That's annoying marketing waste. A false negative means you miss someone who is actually about to leave, so you don't intervene, and they go. That's real business loss.

Which means 93% accuracy in a churn model might be a disaster or perfectly fine, and the accuracy number can't tell you which. If the errors are mostly false negatives, you're bleeding. If they're mostly false positives, you're wasting a little marketing budget. Same metric, opposite verdicts. You measured accuracy, it was high, you declared the model validated, and then in production you're failing at the business problem you set out to solve, because you optimized for the wrong thing.

This scales in every direction that matters. In lending, false negatives (denying credit to someone who would have repaid) cost you revenue, and false positives (approving someone who defaults) cost you losses; the acceptable balance depends on your risk appetite, your business model, your regulatory environment, none of which appear in an F1 score. In hiring, false positives destroy your talent pipeline and burn recruiting spend, while false negatives put people on your teams who won't succeed; the tolerable ratio depends on your training pipeline and your attrition rates and your legal obligations, not on the accuracy metric. Most validation processes never think about any of this. They treat accuracy as a proxy for quality and assume that if the number is good, the decision is good. It frequently isn't.

The Fairness Mirage

It gets worse with fairness, because fairness metrics sound more sophisticated and are actually more fragile.

You've committed to making fair decisions. Good. Now define fairness. *Equal accuracy* (the model predicts defaults equally well across races, genders, ages) sounds right, except that if baseline default rates actually differ across groups, equal accuracy produces unequal impact, and you end up approving proportionally more loans to the group with the lower baseline rate. *Equal impact* (the model rejects the same percentage of applicants from every group) also sounds fair, except that if the groups genuinely have different risk profiles, equal impact forces you into statistical inaccuracy and you start approving loans you shouldn't and rejecting people you should approve. *Equal opportunity* (equalized true positive rates across groups) sounds fair too, except that it assumes non-default is equally distributed across groups, says nothing about false positive rates, and is a different metric than the first two.

There are a dozen-plus competing formal definitions of fairness in circulation, several of them provably incompatible with each other, and no mathematical way to satisfy them all at once. Optimize for one and you fail the others, by construction. Yet most validation processes pick a single metric, measure it carefully, optimize hard for it, and call that "ensuring fairness," then act surprised when the model gets challenged on a fairness dimension they never checked, or when the metric is pristine and the decisions are still systematically wrong for a particular group. The failure is identical to the accuracy failure: you measured a proxy and assumed the proxy was the outcome you cared about. It usually wasn't.

Model Risk Is Not Business Risk

Here is where the boring truth surfaces. A bank's Chief Risk Officer once walked me through a model risk framework. It was comprehensive: data quality checks, concept drift monitoring, backtesting protocols, confidence intervals, stress testing. It looked good. Very academic.

Then I asked: what happens when a model decides to drop a product line that's currently profitable but the model thinks is risky?

Silence.

What if the model's risk assessment is more conservative than the business's risk tolerance, and nobody knows that's the conflict?

More silence.

Who resolves that conflict?

That silence is the gap almost every enterprise is walking into. Most model risk frameworks descend from Basel guidance and financial-regulator requirements, or from research and trading environments where "maximize accuracy" is the entire goal. They are genuinely good at what they were built for: identifying technical failure modes, catching when training data has changed, detecting when model assumptions are violated, flagging when a model's confidence is misplaced. They catch overfitting, concept drift, and data quality issues. These are real and important things. What they do not do (what they *cannot* do, because it's a different category of problem) is tell you whether the decision the model is driving is a decision your business actually wants to make.

Watch how the same failure reads in each world. A research model that fails says: *accuracy dropped from 94% to 91%, something is broken.* An enterprise model that fails says: *the model is rejecting 15% more credit applications than it used to. Is that because credit quality changed, or because we changed the training data, or because the model is malfunctioning, or because the model's risk tolerance diverged from the business's?* The academic framework catches the first two possibilities. It will not catch the one that matters most, because it never asks whether the model's decision-making is aligned with what the business wants. That question isn't in the framework. People assume it's handled by governance instead: "we have a committee that approves what the model does." But the committee approves the model once. It does not continuously validate that the model is still doing what it approved.

So you get the promotion model. A company deploys a model to recommend employees for promotion. The governance committee signs off. Accuracy is good, fairness testing passes, it goes into production. Six months later someone notices it's recommending technical specialists for management roles at far higher rates than the business actually wants. The model learned that technical expertise predicts individual success, which is statistically true, but the business does not want to promote all of its best engineers into management, and that preference was never a variable the model could see. Is this a model risk problem? In the academic sense, no: the model is doing exactly what it was trained to do. In the business sense, absolutely: its decisions conflict with what the business wants. The framework asked *is it accurate, is it fair, is it stable* (yes, yes, yes), and never asked *is it doing what the business wants.*

Real enterprise model risk has three layers, and most frameworks live entirely in the first. **Technical risk:** is the model working as designed (data quality, drift, distribution shift, confidence calibration)? Necessary, and where academic frameworks stop. **Business risk:** is the model's decision-making aligned with business goals (not just accurate predictions, but the actual decisions you want made)? This is the layer that goes unmonitored. **Governance risk:** is someone explicitly accountable for what the model does, and if its behavior conflicts with policy or regulation, do you have a way to know and a way to respond? Monitoring a credit model across all three means watching data quality *and* watching whether a shifting approval rate reflects real credit-quality change or model drift *and* having a named person who has looked at the outcomes and said "yes, this is what we want" — including when the model is declining business that used to be profitable. Most enterprise models are making proxy decisions: you don't actually want a model that's accurate at predicting default, you want one that makes profitable lending decisions, recognizes systemic risk, and complies with regulation. Accuracy is a means to those ends, not the end. A model that's 95% accurate at predicting default while declining 30% of applications that would have been profitable is, to the academic framework, a success. To the business, it's a slow leak.

The Risk That Looks Like Success

The most dangerous failure mode is not the model that breaks. It's the model that is technically sound, makes consistent decisions, passes every monitoring check — and is making decisions the business doesn't actually want. Every model risk framework shows green. The validation is clean. The problem isn't in the model; it's in the decision system around the model, and the framework was never pointed there.

You can have a trustworthy model used for an untrustworthy decision, and it happens constantly. A lending model that accurately predicts default risk produces a disparate impact on a protected class, because it was trained on data from an era when lending practices in that market were different: the model is statistically sound and the decision to use it there was wrong. A trading model that accurately captures market correlations produces an unexpected loss when the correlation structure changes, because nobody ever decided what to do when the model's assumptions broke: the model did what it was trained to do and the decision authority around model uncertainty was never established. A workforce optimization model that accurately predicts which positions will have retention problems gets pointed at layoff recommendations, because nobody decided whether a model built to predict problems should decide who loses a job: the model worked, the application was wrong. In each case the model is fine and the decision context is not, and no amount of accuracy monitoring will surface it.

This is the risk that looks like success, and you won't see it until it matters: until someone asks "why did we decline that credit application?" and the answer is "the model said so," and the question back is "but is that what we actually want?" If you can't answer that quickly and clearly, you have a model risk problem that no academic monitoring will ever catch. The cost, remember, is rarely one bad decision. It's the pattern of decisions that are each individually defensible but collectively add up to exposure you never chose.

Decision Quality: The Question Validation Should Ask

So change the question. Validation should not be checking *does this model work.* It should be checking *does this model produce decisions I'm willing to make and defend.* That's a business question, not a metric, and it forces you to say what you're actually trying to accomplish before you build anything.

The organizations that manage model risk well share one habit: they define decision quality first, then validate against it. That process has four moves. **First, define decision quality as requirements, not metrics.** Not "what's our F1," but "we need to identify 80% of creditworthy applicants without exceeding a 3% default rate," or "we need to catch 60% of likely churners and we'll fund interventions that succeed half the time," or "we need candidates with a greater-than-70% two-year success rate and at least 30% representation from underrepresented groups in the hire cohort." These are business requirements, and a data scientist cannot write them for you. **Second, measure whether you're meeting them.** This is where metrics finally earn their place, but they're metrics aligned to the requirement: not "what's our fairness score" but "do we have the representation we said we wanted, and at what cost to accuracy." **Third, make the trade-offs explicit.** Every model trades something for something (accuracy for fairness, profit for customer risk, speed for accuracy), and in most organizations these trades just happen, silently. Build validation that says, in concrete terms, "if we deploy this we gain X in business value and accept Y in risk, and here is exactly what Y looks like," and make sure the person approving the deployment actually understands what they're approving. **Fourth, set the failure threshold before deployment.** Not "we'll know it when we see it," but "if this metric crosses this line, we reconfigure or we pull it." You may not care if accuracy drifts from 93% to 90%. You almost certainly care if the model starts violating your fairness requirement, and you definitely care if the cost per intervention rises past the point where the program stops making economic sense.

The Unglamorous Work That Actually Catches It

Defining decision quality up front tells you what "good" means. It does not tell you when reality has drifted away from it — and reality always drifts. The system gets deployed. It works. The engineers are happy because the metrics hold, the business is happy because decisions are faster, everyone feels good. Then, slowly, a gap opens between how the system is used and how it was designed to be used. A human starts trusting it too much and stops applying judgment. Someone uses it in cases it was never built for. The process changes and the model doesn't. The data drifts because the business shifted and nobody told the model. None of these are failures. They're just reality: real humans, real workflows, real organizations, where things change, people adapt, and systems sometimes don't. In the gap between how the system was designed and how it's actually being used, risk accumulates silently.

You cannot fully prevent this. But you can dramatically reduce it, and the fix has one prerequisite most organizations skip. Chapter 12 works through what continuous monitoring should look like as a discipline; the piece that belongs to *validation* is narrower and very specific. Log the human's decision alongside the model's prediction, on every case. Not the prediction alone — the pair. Do that and the gap between what the model recommended and what the organization actually did stops being an anecdote and becomes a measurable object, and you can finally ask the questions that test decision quality rather than model health. When the model says decline and an underwriter overrides to approve, how often does that borrower repay? When the model says approve and a human declines anyway, how often would it have been fine? Is the override rate rising, falling, or concentrated in one product, one region, one shift, one person? An override rate that slides from 12% to 2% is not evidence the model got better. It usually means the humans stopped reading: the exact failure that was accumulating in the lending case above, and one that nothing on the model side will ever surface, because on the model side nothing changed. And then the second half of the job, which is the least glamorous control in this book: an actual human with authority reading actual cases, regularly, manually, asking *does this feel right, where should I worry.*

The lending organization from the start of this chapter, once they understood what had happened, did exactly this. They assigned a risk officer to read 50 random decisions a week, just reading them, just asking whether the approvals made sense and matched the firm's risk appetite. That single investment, one person's time, would have caught the drift before it became millions in tail risk. Most organizations won't do it. It feels expensive, it feels like overhead, the system is working and the metrics are good, so why pay a human to read cases? Because operational risk isn't in the metrics. It's in the gap between what you're measuring and what's actually happening, and you cannot see that gap from a dashboard. You can only see it by looking at real decisions, real outcomes, real patterns of behavior in the world.

The organizations that separate themselves from the pretenders do one more hard thing beside the framework. Call it the uncomfortability conversation. Someone sits down and lists every way the model could produce a bad outcome the framework won't catch: accurate prediction of biased historical patterns (model correct, decision wrong), high accuracy but low coverage (works for some populations, not others), decision-context change (trained in one business context, deployed in another), correlation breakdown (captures historical relationships that don't hold going forward), population drift (trained on one population, applied to another). For each one they ask the only three questions that matter (*do we have a process to catch this, is it monitored, would anyone actually know if it happened*), and the answer is usually no. Then they do the hard part: they either redesign the model and the decision system so the real risks are visible and manageable, or they decide the risk is too high and don't deploy.

Five questions do most of the work in that room, and they're worth asking, in order. *What is this model actually being used for*: not in theory, in practice; how is the decision really made, and who is on the receiving end of it? *Why are we using a model for this decision at all*: because it's more accurate than the humans were, because it's faster, or because it launders the appearance of bias out of data that is still biased? Those are three different risk profiles wearing the same deployment. *What's the decision if the model is wrong*: when confidence is low or the prediction is bad, what happens next, concretely; if you can't describe that, you don't understand your own exposure. *What would falsify our assumptions*: you are assuming the training data represents the population you're applying it to and that the decision context is stable, so name the thing that would prove either one wrong. *Is there a human who could reasonably override it*: not should there be, could there be; if the honest answer is no, the model is making the decision, and you should say that out loud rather than keep a human in the org chart for comfort. The move that matters is to push on the answers rather than accept them. If someone says "the framework handles it," ask how, exactly: describe the alert, describe the decision that gets made when it fires. If someone says "human judgment handles it," ask what training those humans have and how you'd know it's working. If you can't get a satisfying answer to either, the model isn't ready to deploy, not because it's mathematically broken, but because you don't understand the risks it creates in your actual business.

This is the boring work, and it won't show up in quarterly metrics. It also happens to be the whole game. When something eventually goes wrong (and with a system making real decisions at scale, something eventually does), someone will ask you to justify the deployment. *This model made this decision. Why did you think that was acceptable?* Your answer cannot be "our F1 score was 0.87." It has to be "we required this outcome, we validated that we achieved it, we understood the consequences, and we decided it was acceptable." One of those answers you can defend in a regulatory meeting. The other one is how you end up in the regulatory meeting. Your validation process was never just a technical gate. It's the evidence for your governance decision, so make sure it's measuring what actually matters.

FIGURE 6.1

The Silent-Failure Quadrant

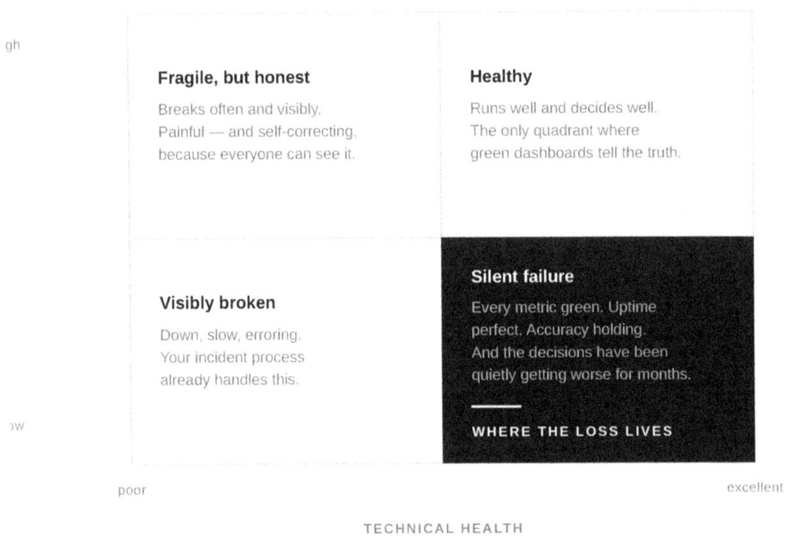

Workshop use: teams plot their live models and defend each placement. Anything landing in the dark quadrant becomes an immediate review item.

FIGURE 6.1

The Bottom Line

Accuracy tells you a model works; it says nothing about whether deploying it is a good business decision, and the whole industry has built validation processes that answer the first question and assume they've answered the second. A model can be 95% accurate and still bleed money, because the cost of a false positive isn't the cost of a false negative, because the competing definitions of fairness are provably incompatible, so optimizing one guarantees failing others, and because a model trained to predict accurately gets pointed at decisions it was never designed to make: promotions, layoffs, lending in markets its training data misrepresents. The risk that actually costs you is the one that looks like success: green on every check, drifting silently in the gap between how the system was designed and how it's really used. The fix is mostly non-technical: define decision-quality requirements before you build, monitor decisions and not just the model, put a human with authority in front of real cases every week, and have the uncomfortable conversation about how this thing fails before it does. When something goes wrong, "our metrics were good" is not a defensible answer; "we required X, validated it, understood the trade-off, and chose to accept it" is. Building the second answer is what validation is actually for.

Field Kit

1. The error-cost worksheet (bring before you approve any model). For every model driving a consequential decision, write down the two error types in business terms and price them: what does a false positive cost in dollars, what does a false negative cost, and which one are you currently optimizing to avoid? If the two costs aren't roughly equal — they almost never are — a single accuracy number is hiding your real exposure. No model advances past this step on accuracy alone.

2. The decision-quality requirement, written before the model is built. State the requirement as a business outcome, not a metric: "identify 80% of creditworthy applicants without exceeding a 3% default rate," "catch 60% of likely churners with interventions that succeed 50% of the time." One sentence, owned by the business, not the data science team. If you can only express what you need as "high accuracy," you haven't defined the requirement yet.

3. The three-layer monitoring check. For each model in production, confirm you are watching all three layers, not just the first: **Technical** (data quality, drift, distribution shift), **Business** (are the actual decisions aligned with the outcome you wanted, and is any change in behavior intentional), **Governance** (is a named person accountable, and is there an escalation path when the model's decisions diverge from policy). Most organizations can only answer for the technical layer — that gap is the finding.

4. The five pre-deployment questions. Ask these in the room, out loud, before anything ships: What is this model actually being used for, in practice? Why are we using a model for this decision at all? What's the decision if the model is wrong? What would falsify our assumptions? Is there a human who could reasonably override it — not should there be, could there be? If any answer is vague, push once — "describe the exact alert, or the exact decision when it fails" — and if the second answer is also vague, the model isn't ready.

5. The weekly decision read. Name one person with authority to read a fixed sample of real decisions every week — 50 is a defensible starting number — and give them one job: does this batch make sense, and does it match our risk appetite? Log the human decision alongside the model's prediction so the gap between them is visible and trending. This is the single cheapest control that catches the risk metrics can't see.

6. The uncomfortability conversation, written down. List the five silent failure modes — accurate prediction of biased history, high accuracy / low coverage, decision-context change, correlation breakdown, population drift — and for each answer three columns: Do we have a process to catch this? Is it monitored? Would anyone know if it happened? Any row that's "no / no / no" is either a redesign task or a reason not to deploy. Keep the register with the model and revisit it whenever the business context shifts.

CHAPTER 7

Model Risk in the Foundation-Model Age

Part III · Model & Data Risk

The head of data science at a mid-size insurance carrier once showed me the memo that chose the model her company would use to triage customer feedback. It ran to one page. Model A: 87% on the standard benchmark. Model B: 84%. Recommendation: Model A. Three points, one signature, done. I want to be fair to her — she was a careful person doing exactly what her governance framework asked, and I've read versions of that memo, in different fonts, at a dozen companies.

Six weeks after deployment she was in a conference room explaining why the numbers hadn't predicted reality. Model A wasn't catastrophically worse than Model B would have been. Just noticeably, persistently worse, on a system that had cleared every gate the company had.

Here is the reframe this chapter is built on.

The myth: model risk is a validation problem, where you hold the model still, understand it, measure it against a benchmark, and sign it off before deployment.

The reality: foundation models cannot be held still or understood, so governance shifts from explaining the model to continuously watching how it behaves on your data, in your context.

This is where model risk actually lives: in the gap between the benchmark and production. And that gap isn't a measurement problem you can close with a better test. It's a symptom of something larger: almost every enterprise is managing foundation models with a model-risk apparatus built for a different kind of model entirely. Most organizations inherited their model-risk framework from financial services, and usually from one document: the supervisory guidance on model risk management issued by the Federal Reserve and the OCC in 2011, known to everyone who works with it as SR 11-7. It is a genuinely good piece of regulatory writing, and it was designed for regression models that predict loan default or market behavior.

Those frameworks also assumed the data beneath the model was settled, which Chapter 8 takes apart. But their central bet was narrower: that you could hold the model still and understand it. Extract the coefficients. Measure the bias. Validate the performance. Sign off. Deploy. The model's behavior was relatively predictable within its operational envelope, and if you wanted to know why it did something, a data scientist or a mathematician could, in principle, crack it open and trace the causal path from input to output. Tedious and expensive, but possible.

That world no longer exists. And the governance you built for it is now pointed at the wrong problem.

The Benchmark Number Means Less Than You Think

Start with the model selection decision, because it's the first place the old assumptions fail and almost nobody notices.

A benchmark measures one specific thing: how well a model performs on well-formed inputs that look like its training data. The test set is clean (invalid inputs removed). It's labeled (the right answer is known). It's representative of the problems the model claims to solve, and the model sees it under controlled conditions. That's genuinely useful information. It's also a narrow slice of what you need to know. The benchmark doesn't tell you how the model handles ambiguous or contradictory inputs, how performance degrades as the input distribution shifts, what happens when the model is confident and wrong, how behavior varies across user segments, or what the latency-accuracy trade-off looks like when you need faster inference. All of those things matter in production. Benchmarks typically measure none of them.

The reason is structural, not sloppy. Benchmarks are optimized for comparability, not prediction. The entire point is a standardized, reproducible test scored against one test set with one objective truth. Production is the opposite: shifting input distributions as your users' problems change, ambiguous ground truth where sometimes there's no right answer, diverse use cases as the same model gets pointed at different purposes by different teams, and a real, unequal cost of being wrong, because the benchmark weights every mistake the same and your business doesn't.

Here's how that played out at the carrier. The benchmark for sentiment models tests on a curated dataset of reviews where the sentiment is unambiguous, and Model A scored 87% to Model B's 84%. So she picked A. But what actually arrived in production was short support-ticket messages, claims correspondence, product reviews, social mentions, conversational text. Some of it was ambiguous: is "I'm shocked at how fast this is" positive or negative? Some was sarcastic. Some was context-dependent, where a complaint to a friend reads nothing like a bug report. The benchmark tested none of that. Model A, optimized for the benchmark, was more confident on the ambiguous cases: it handed the triage queue clean answers and told it to move on. Model B, three points lower, had more calibrated uncertainty; it flagged the ambiguous cases, which is exactly the signal telling you where to apply human judgment. In production, Model A's apparent superiority became a liability (more confident wrong answers, more escalations that nobody saw coming), while Model B's caution, the property that cost it three benchmark points, was the more valuable one.

None of this is an argument against benchmarks. It's an argument against making the benchmark your only signal. The enterprises that select models well run a longer, more boring process. They start with benchmarks to eliminate the obviously bad choices. Then they do the step almost everyone skips: they run both finalists on a sample of their own messy, actual inputs (not the benchmark test set) and watch the rankings change. They evaluate on their own criteria, because the question for their use case might not be accuracy at all but calibration, latency, or a specific pattern of mistakes they can tolerate. They look at failure modes: does one model fail gracefully by flagging uncertainty while the other fails confidently and wrong? And only then do they run a small production pilot on real traffic to see whether their intuitions about risk held.

Step two, test on your own data, is where most enterprises stop short. They run the benchmark, pick the winner, and assume the work is done. It isn't. They've eliminated the worst choices; they haven't eliminated the wrong choice for their specific situation. So treat the benchmark as a starting point, not a conclusion, and ask the vendor the questions it can't answer (about ambiguous inputs, distribution shift, confidence calibration, and whether the model fails confidently or uncertainly). The benchmark number is real. It's just not as predictive as it looks.

You Cannot Understand the Model, and That Is Not Temporary

The selection problem is a preview of a deeper one. There's now a category of risk that your model-risk framework was never written to contemplate: deploying systems whose internal workings you cannot meaningfully understand, even if you built them.

Traditional machine learning was comprehensible in a specific way. A random forest combined the decisions of many trees. A linear model had coefficients you could read. A neural network had weights and activations you could instrument. All complex, but the complexity was a matter of scale and dimensionality, not fundamental incomprehensibility. With enough time and expertise, someone could explain the path from input to output. Foundation models are a different thing. Trained on so much data, using such high-dimensional representations, that their internal decision-making is genuinely opaque. You can measure what a foundation model produces. You can test it. You can't look at it and explain why it produced a particular output in a particular context, in the way governance frameworks expect.

People call this the black box problem, but that frame is misleading. It isn't that the box is black; it's that there's no meaningful box at all, just a high-dimensional space of patterns learned from data, activated in context-specific ways that don't decompose neatly into anything a human can hold. And this matters for one uncomfortable reason: complete interpretability of foundation models at this scale is, for now, an unsolved problem — not a temporary tooling gap you can wait out. Current approaches to understanding a 70-billion-parameter model run into limits that reflect deeper constraints around human cognition, time, and the sheer dimensionality of these systems.

This is the myth most leaders are still betting on: *We'll use SHAP. We'll use attention visualization. Interpretability research is progressing, and once it catches up, our governance problem is solved.* That research is valuable and progressing, and higher-stakes deployments genuinely warrant more of it, but building your governance on the assumption that full interpretability arrives before you have to act is building on a foundation that may never pour. The best-run organizations I work with have stopped waiting. They've made the move this entire chapter turns on: from *understand the model* to *understand its behavior in your context.*

That's a completely different question, and the good news is that it's answerable. You may never know why the model produces a given output. But you can know how it behaves across user populations, where it fails systematically, which edge cases make it unreliable, and where it makes decisions that surprise you. This is behavioral understanding rather than mechanical understanding, and unlike mechanical understanding, it's actually achievable with systems you can't fully explain. Nobody gets to present a beautiful causal diagram of the model's reasoning; you get instrumentation, dashboards, and someone whose job is to watch for surprises. That's the trade the foundation-model age forces on you: give up the fantasy of understanding the machine in exchange for the discipline of watching it.

Accepting that constraint forces three shifts. You move from internal validation (analyzing the model's internals) to external validation: observing behavior in the real world, deploying and monitoring and measuring whether the model actually creates the value you expected. It's slower than achieving complete understanding up front, but it validates against ground truth rather than your incomplete mental model. You move from designer intent to observed outcomes, because a foundation model may be excellent at the task you tested and terrible at one someone found in production; governance becomes about what the model actually does, not what you meant it to do. And you move from point-in-time validation to continuous validation, because behavior drifts and changes as people discover new uses: validation stops being a gate you pass and becomes an ongoing process.

It Is Not Model Risk. It Is System Risk.

Even behavioral understanding of a single model isn't enough, because you're almost never deploying a single model. You're stacking them. An LLM calls an API that reads a database, transforms the data, feeds it into another model, and hands a formatted answer to a human who acts on it. The risk surface is no longer one model but a web of interconnected systems, each with its own risk profile, each capable of amplifying or masking the others' failures. That's not model risk. That's system risk, and a framework built to validate one model in isolation is built for the wrong unit of analysis.

Three dimensions of complexity are arriving faster than most organizations realize, and each one breaks a tool the old framework depended on. The first is **behavioral drift.** Your model performs differently in production than in test (a known problem), but with foundation models the drift becomes harder to detect and easier to miss. You're monitoring for deviation from expected performance, but what's "expected" when the model's output distribution is continuously shifting based on user patterns, retrieval data, and context? The drift-detection frameworks that worked in traditional ML assumed the thing drifting was small enough to hold in your explanation budget. Now it isn't.

The second is **emergent capabilities.** You deploy a model for Task A, and in production users discover it's useful for Tasks B, C, and D (sometimes better than your original use case). Those capabilities were latent in the model. You didn't design them, you didn't test them, and you don't fully control when they appear — but you're now accountable for them. A governance framework that only reasons about the capabilities you architected is silent on exactly the behaviors most likely to surprise you.

The third is **compounding amplification.** When you stack systems, failures don't merely accumulate. They amplify. A hallucination in one model becomes the input to another, which makes a high-confidence, wrong decision based on false data, which flows downstream as if it were a fact. Your risk assessment for each individual model can show green while the composed system is dangerous, because nobody is reasoning about system-level failure modes.

Set against these three forces, the model-risk programs I see even in mature enterprises look thin. They usually have three components (testing and validation, performance monitoring, and escalation procedures), and within their scope they measure the right things. The problem is scope. What's missing is behavioral monitoring in context: observing how the model behaves at the boundaries of its training data, with adversarial input, with out-of-distribution populations, because that's where real risk lives. What's missing is feedback-loop integration: monitoring tells you the model is drifting, governance escalates to data science, someone fixes it, and nothing you believed about the model changes as a result. And what's missing is stack-level risk assessment: your model is accurate, your retrieval system is reliable, your API is stable, and the combination of the three, with a human relying on the output, has a risk profile none of the component reviews captured.

When AI Governs AI

Push the scaling problem one more step and it turns recursive. A platform lead at a payments company walked me through his inventory once: forty-odd models in production, another dozen in shadow mode, and a two-person team fielding every anomaly alert the whole fleet generated. He wasn't asking whether to automate the triage. He was asking what he was supposed to tell his risk committee when he did.

That's the shape of the problem. Today a human reviews the anomaly alert, decides whether to investigate, retrain, or pull, and is accountable for the call. At fifty systems that becomes untenable (you cannot have a human review every anomaly for every model), so you build a system to monitor the fleet and either recommend actions or take them directly, escalating the edge cases. And now you have a genuinely new governance problem: how do you govern the AI that makes decisions about AI?

Recursive governance has counterintuitive properties. The feedback loop is worse: when a human makes a bad call, you notice, correct it, and the human learns; when an automated decision system makes a bad call, it can repeat the same mistake across the fleet before anyone notices, and a bad decision about one model can trigger bad decisions about others. Accountability blurs: when a human is wrong, the responsible person is obvious; when the system is wrong, you're suddenly asking who built it, who owns its training data, who monitors its decisions — the kind of question that lands on the desks of your CRO, CTO, and GC at once, with none of them sure it's theirs. And complexity multiplies, because you're no longer just governing the base models; you're governing the decision-making system, and then the governance of that system.

The payments team's first build showed exactly how subtle the failure is. Their orchestrator triggered retraining when a model's accuracy dropped below a threshold, then checked whether accuracy improved and treated "yes" as success. That system was learning — but learning what? That retraining works? Or that the retraining process was fundamentally broken for a class of models and it should stop trying? Two quarters in, it was cheerfully retraining a set of models on a monthly cadence that never needed it, and leaving alone a handful whose underlying data had shifted so far that retraining couldn't have helped. The metric it optimized looked fine the entire time.

The way through is to treat the decision-making system as exactly what it is: a critical production system that needs governance as much as any model it oversees, applied one level up. That means interpretable logic you can challenge for a call like "pull this model": *I recommended pulling because accuracy dropped X%, monitoring shows Y, confidence is Z,* not a black box making decisions about black boxes. It means an explicit line between decisions that can be fully automated, decisions that need human review, and decisions a human must make; monitoring the decision system for its own biases (toward pulling, toward conservatism, toward always retraining); audit trails so that when a decision about Model A affects Model B affects Model C you can trace the cascade; and deliberate, structured feedback rather than automatic learning, because a governance system that rewrites its own rules on the fly is a faster way to compound errors, not a solution. Start conservative (narrow scope, high transparency, most decisions escalated) and expand automation only as you earn confidence. Governance itself, it turns out, needs to be governed.

What Actually Works

Strip away the sophistication and the recursion, and the governance that works in the foundation-model age comes down to a handful of disciplines, applied relentlessly. None of them require you to understand the model. All of them require you to watch it.

Have your governance framework reason about the entire decision path (data ingestion, inference, output formatting, integration with business systems, and the human decision at the end), because a failure anywhere in that path is your problem, regardless of which component technically broke. Shift from static testing to continuous behavioral observability: instrumentation that tells you not whether the model is accurate on average, but where it fails systematically (which populations get different outputs, where confidence is high while accuracy is low, where edge cases are emerging). You can't enumerate every failure mode of a billion-parameter model, so monitor for the ones that would hurt most: whether outputs stay coherent, align with your data, and make sense in context.

Make your assumptions explicit and then check them continuously. Every deployment carries assumptions: this model works well with data like X, it's safe for use case Y, it behaves for populations like Z. Write them down, decide in advance when they'll be tested, and build the signals that tell you when they start to break. Separate risk tolerance by use case, because a model that informs a decision has a different risk profile than one that makes the decision, and a model affecting one customer has different requirements than one affecting millions. One gate for everything isn't granular enough. And have escalation protocols ready before you need them: when something surprises you, do you limit deployment, investigate, pull in human feedback? You can't treat every surprise as a crisis, but the structured way to decide what deserves escalation should be settled in the calm before, not improvised in the panic during.

Which leaves the discipline nobody puts in a deck: what happens *after* the escalation. Chapter 4 made the general case for closing feedback loops; the foundation-model version is narrower and more mechanical. Your escalation path has to end in a changed line in the assumption register — not a closed ticket.

Go back to the carrier. Nine months after go-live, the sentiment model started mis-scoring a whole category of messages: customers who'd moved to the new mobile app wrote in a clipped, abbreviated register the model read as curt and negative, and satisfaction scores for that channel sank for reasons nobody could see. The team handled it well by conventional standards. They caught it, diagnosed it, retrained, and watched the metric recover. Ticket closed. What they didn't do was write down the sentence that mattered (*this model degrades whenever a new channel changes how customers write*) or attach a signal that would fire the next time it happened. So when the carrier launched an SMS claims channel the following year, they rediscovered the identical failure from scratch and paid for the discovery a second time.

That's the whole distinction. Remediation restores the metric. Governance changes what you believe about where the model can safely operate, and records it somewhere the next deployment will trip over it. If a model drifted, you should come out the other side knowing why and having updated the register, not just having fixed the number. The work is entirely unglamorous: a meeting, a document, someone maintaining a list of assumptions that nobody reads until the week they urgently need it. It'll never show up in a quarterly metric. It's also the whole game.

None of this is perfect governance. It's governance in the age of systems you can't fully understand — the only age you're actually operating in. The alternative, deploying black boxes and trusting that explainability research or a clean benchmark has you covered, isn't governance; it's hoping for the best at a scale where your board should be deeply skeptical of hope as a strategy. The complexity is coming whether you're ready or not. The only real question is whether you'll understand what your models are doing when it arrives.

The Bottom Line

The model-risk framework you inherited from financial services assumed you could hold a model still and understand it. Foundation models have voided every clause of that assumption. You can't fully interpret a 70-billion-parameter model, and that isn't a tooling gap you can wait out. So stop trying to understand the model and start understanding its behavior in your context: how it performs on your messy data rather than a clean benchmark, where it fails systematically, and whether the assumptions you deployed it under still hold. The unit of risk has also moved. Stacked models, emergent capabilities, and compounding amplification produce failures no single-model review will catch, and the decision systems you build to watch your fleet are production systems that need watching too. What works isn't sophisticated. It's boring and continuous: expand "model" to mean the whole decision path, replace point-in-time validation with behavioral observability, monitor your assumptions, tier risk tolerance by use case, and make every escalation end in an updated assumption rather than a closed ticket. Perfect understanding was never on offer. Disciplined visibility is.

Field Kit

1. The procurement question set (bring to every model selection). Before you let a benchmark decide, ask the vendor the four questions benchmarks can't answer: How does this model perform on ambiguous or out-of-distribution inputs? What happens to accuracy as the input distribution shifts? How calibrated are its confidence scores? When it fails, does it fail confidently or uncertainly? Then do the step everyone skips — run the finalists on a sample of your own real, messy inputs and see whether the benchmark ranking survives contact with your data. A three-point benchmark gap that comes from confidence rather than correctness isn't a decisive advantage.

2. The "understand the behavior, not the model" checklist. For any foundation model in production, confirm you can answer five behavioral questions without needing to explain the model's internals: How does it behave across different user populations? Where does it fail systematically? Which edge cases make it unreliable? Where is confidence high but accuracy low? Where is it making decisions that surprise you? If you can't answer these, you have an instrumentation gap, not an interpretability problem — and instrumentation is the one you can actually close.

3. The assumption register. Write down, per deployment, exactly what you're assuming: the data it works well with, the use cases it's safe for, the populations it behaves for. For each assumption, record when it will be tested and what signal tells you it has broken. Revisit the register on a schedule and whenever the business context shifts. A model without a documented, monitored set of assumption boundaries is a model you're trusting on faith.

4. The system-risk map (not the model-risk review). Diagram the full decision path — data ingestion, model inference, any stacked models or APIs, output formatting, business-system integration, and the human decision at the end. For each hop, note how a failure there propagates downstream, with special attention to where a wrong-but-confident output from one component becomes trusted input to the next. The review isn't complete until someone has assessed the composed system, not just the components.

5. The use-case risk-tolerance tier. Classify every deployment before it ships: does the model *inform* a decision or *make* it, and does it affect one customer, thousands, or millions? Assign the governance intensity — monitoring frequency, escalation thresholds, human-review requirements — to the tier, not to models uniformly. Higher-stakes tiers warrant more rigorous interpretability effort and tighter escalation; lower-stakes tiers shouldn't drown in the same overhead.

6. The closed-loop escalation protocol. Decide, before you need it, what happens when a model surprises you: the criteria for limiting deployment, for investigating, for pulling in human feedback, and for pulling the model. Then add the step most organizations omit — after remediation, require a written answer to "why did this happen and what assumption do we now update?" and confirm that the answer lands in the register with a signal attached. An escalation path that ends in remediation is maintenance; one that ends in an updated assumption is governance.

CHAPTER 8

The Data Foundation: Quality, Provenance & Meaning

Part III · Model & Data Risk

"Our data quality isn't good enough." Someone says it in every enterprise AI deployment, usually early, usually with folded arms, and what they mean is specific: *I've looked at the source systems. Some fields are empty. Some are wrong. Some mean different things depending on which department entered them.* Everyone nods solemnly, as if this were a fresh and troubling discovery, as if 1980 called and wants its data problems back.

The lie isn't that you have bad data. Of course you do. The lie is the myth of *good enough*.

Most organizations operate under a belief that's almost never stated but sits in the room like furniture: if you can get your data to 90% quality, you can build a functional AI system. Maybe 85%. Definitely at 95%. This is false. Data quality matters, but you can build a functional AI system on pretty broken data. You just can't build one you can trust to make decisions at scale.

Here is the reframe this chapter is built on.

The myth: the model is the primary risk surface, so AI governance belongs around model cards, testing, and deployment gates.

The reality: the model is only an amplifier; the real risk surface is the data, and governance fails because you cannot prove where it came from or agree on what it means.

You've structured your AI governance around the *model* (model cards, risk frameworks, monitoring, explainability, deployment gates) on the assumption that the model is the primary risk surface. It isn't. The model is an amplifier. The actual risk surface is the data. Biased training data, and the model amplifies the bias; missing edge cases, and it can't handle them; gaps in collection, and the model has blind spots that map to them exactly. Every model risk problem I've seen at enterprise scale had a data problem at its root. The model was fine — learning precisely what it was trained to learn. What it was trained to learn wasn't what anyone thought.

And governance fails at the data layer for two reasons almost no framework is built to catch: enterprises can't prove where their data came from, and they can't agree on what it means. Everything in this chapter is a variation on those two failures.

The Myth of Good Enough

Start with quality, because it's where everyone starts and where everyone stops too soon.

I've watched teams spend eight months fixing data pipelines. Getting the nulls down. Standardizing schemas. Running reconciliation checks. They get the system into production and it performs fine on the metrics. But when a human looks at what it's actually recommending (especially in the edge cases, the long tail), there's something *off* about it. Not catastrophically. Not always. Just off. The system learned the pattern, including the pattern of your data quality problem.

It helps to be precise, because your data has three kinds of defect and they are not equally dangerous. **Structural defects** are the easy ones: missing values, wrong types, things that break a schema. Your tools catch them, you fix them, life goes on. **Semantic defects** are harder: a field labeled "revenue" that means net in one business unit and gross in another; a "customer type" coded one way three years ago and differently now, with the history never updated; a date field that's sometimes a transaction date and sometimes a creation date depending on which system it came from. These are invisible to structural checks. They look clean on a scorecard. And they teach your model nonsense. **Temporal defects** are the ones that kill you. Your historical data is, by definition, old: collected under different business conditions, different processes, different market dynamics. A system that learns from it is learning from a past that no longer exists. You discover this about six weeks after go-live, when the predictions start diverging from reality.

So the real question isn't *is our data good enough?* It's *what decisions are we letting this system make, and which defects can we actually live with?* This is where conversations break down, because it forces you to be specific. And specific is uncomfortable. A recommendation engine that occasionally suggests a slightly irrelevant product? You can live with that; the customer ignores it. Cost of failure:

low. A system that scores credit risk or flags fraud and has learned to be structurally inconsistent about what "fraud" even means? You can't live with that. You'll catch it in compliance review three months in and rebuild it.

The practical move isn't to announce you need "better data quality." That's a budget request wearing a governance costume, and it has no finish line. It's to map your specific defects against the specific decisions you're making, then deliberately reduce the scope of decisions you let the system make until the defects stop mattering. Keep it advisory instead of binary. Restrict it to the business units whose data you understand. Run it in shadow mode longer than you planned, because the temporal drift is real.

This is boring. It won't make the quarterly exec summary. But it's the difference between a system that hums along for three years and one that starts contradicting its own logic in month eight because two semantic defects taught it two incompatible patterns. Your data quality problem was never that you need better data. It's that you need specific knowledge of where bad data actually matters, and the organizational discipline to say: *not yet, not there, not until we understand this.* That's harder than buying a data quality tool — and it's exactly the kind of hard that separates deployments that compound value from the ones that become cautionary tales.

Governance That Stops at the Model

Here's the fastest way to spot a governance framework that won't hold. Ask one question (*"What happens when your training data is wrong?"*) and watch how long it takes anyone to answer.

Most AI governance lives entirely downstream: model cards, testing protocols, access controls, deployment gates. Real things, but every one assumes the data feeding the system is adequate, and that assumption is almost never examined. I've watched it fail in three predictable shapes. The first is the organization with excellent data quality standards — for its transaction systems; when the AI team pulls data for training, it's treated as a novelty, with no service-level agreement, no validation pipeline, no shared definition of "good." The second built its data governance years ago around regulatory reporting, so its lineage tools work beautifully for financial statements but were never designed for what AI training pulls from: third-party datasets, unstructured feedback, behavioral logs. The third keeps the two disciplines in separate houses: data governance owned by engineering, AI governance owned by compliance, emailing each other occasionally while each optimizes locally.

The ending is the same in all three. The system goes to production. Three months in, a downstream process starts failing in ways that don't quite make sense. You dig. The training set included records from a system migration that should have been flagged as an outlier period. Or someone changed how a critical field was calculated and never documented it, so your models run on a stale definition. Or 15% of the training set turns out to be duplicated because a batch job ran twice. By the time you notice, decisions are stacked on top of the model, and your framework has no way to retroactively assess what changed, because it never had a real view of what the data actually was. The question *"Should we have caught this?"* has no answer.

Proper data governance for AI is concrete. Your data contracts extend upstream to the source: not just what fields exist, but what they mean, how they're calculated, when they change. That is harder than it sounds. A "date" field looks trivial until you find it has been calculated three different ways in three legacy systems that were supposed to be consolidated years ago and weren't. You bring a testing mindset to data that matches your testing mindset for code: before a dataset feeds a production model, it passes automated checks for unexpected nulls, distribution shift, and unfamiliar values. Automated, running every load, not a checklist someone ticks once at onboarding. You carry traceability past "we trained this model on this dataset" to "this field came from this system, was transformed by this pipeline, and cleared these validations at each step," because when something breaks you need the whole lineage in an afternoon, not a quarter. And each critical dataset has an owner in the sense Chapter 1 established. The data-layer addition is what that name is empowered to stop: when the owner says "this source isn't ready," that is a model-deployment blocker carrying the same weight as a failed model validation, and it lands before training starts, not after the model is in front of customers. The organizations that do this well don't separate AI governance from data governance at all. The criteria for putting data into production are the same criteria for putting a model into production: same rigor, same review, same ownership. That alignment is harder than any framework, and it's the whole difference between governance that catches problems and one that checks a box.

Where Did This Come From?

Ask a simple thing of any model in production: *"What data did this train on?"* Most large organizations cannot answer with any specificity. That's not an audit inconvenience — it's a blocker to understanding your own model's behavior, managing its risk, and defending its decisions.

Picture a scenario that happens far more often than anyone discusses in public. Your lending model makes a recommendation that looks wrong for a specific demographic segment. Mathematically it's sound, and overall the model performs as expected, but in this segment something is off. Your first question should be: *what data did the model train on for this segment?* If you know ("47,000 approved loans in this geography from 2018 to 2020, and here's the selection bias baked into that set"), you can investigate. Maybe the bias is the problem. Maybe the economic conditions of that window no longer hold. Maybe lending practices have changed. If you *don't* know, you're stuck. Accuracy metrics look fine. Feature importance shows the expected patterns. You can stare at the model all day and conclude nothing is wrong, even though something clearly is, because you have no idea where its learned patterns came from. Organizations that hit that wall usually do one of three things, none good: strip the suspicious feature and retrain and hope; bolt on an override rule and maintain the patchwork forever; or document the concern, ship anyway, and bet auditors have bigger fish. All three are symptoms of the same disease: not understanding your training data.

Provenance tends to break in three places. **Collection and compilation:** you're training on internal transaction history, third-party enrichment, and purchased datasets merged in a pipeline, and years later "what exactly went into the training set?" means hunting through old notebooks, pinging data engineers who've changed teams, and squinting at batch jobs rerun dozens of times with different parameters. What you want to know is simple; what you can reconstruct is murky and partial. **Transformation:** the raw data gets cleaned, features engineered, outliers handled, all in code that evolves, so one version produced the training data and a later version handles new data differently, and when the model misbehaves you can't tell whether the world changed or your pipeline did. **Consent and licensing**, the most dangerous: you've assembled data from sources with usage restrictions and trained on data you shouldn't have had, or used it in a way the agreement didn't permit. That one is nearly impossible to unwind: the model is in production, the restricted data is blended in, and you can't untrain what it learned. You either acknowledge the legal exposure or hope nobody notices.

The fix isn't heroics; it's treating training data as a versioned artifact from the start. Practically, you can answer three questions cold. *What data trained this model?* Specify a version or date range, get back a manifest. *Has that data been modified?* You can tell whether it was processed differently at training time than now, and reproduce the historical processing, because you versioned the feature-engineering code the same way you versioned the model. *What are the licensing and consent restrictions?* Established at the point of procurement, not discovered in a courtroom. The payoff: you can link provenance to behavior and make an argument instead of a shrug: "This model learned from 2018–2020 data in this geography; here's what conditions looked like then, here's now, and here's where we'd expect it to diverge." Regulators are increasingly blunt about it: *your model recommended X for this segment; tell us about the training data for that segment and why the recommendation is fair.* If you can't answer, you don't have a fairness problem yet. You have a provenance problem, and it's about to become a fairness problem in front of an audience.

What Does This Even Mean?

A data governance team came to me frustrated. A compliance audit had flagged undefined data quality standards, which stung, because on paper they were doing everything. They tracked classification: PII, confidential, the rest. They had retention policies. They knew where their data came from. What they couldn't do was explain what it actually *meant*. They had a customer ID field but no definition of which customers it represented: everyone who ever registered, or only active accounts? They had a "revenue" field with no ruling on gross versus net, or transaction time versus recognition time.

This wasn't a gotcha. It's the foundation of data governance, and almost everyone skips it. The fix is a data dictionary: a document, ideally in version control, that defines each data element you use for AI, says what it means, notes any transformation applied, and states the quality you expect. For one field it reads like a single paragraph:

> ***Field name:*** `customer_acquisition_date` ***Definition:*** *The date the customer created an account in the system.* ***Data type:*** *DATE* ***Source:*** `accounts.created_at` ***Notes:*** *Included internal test accounts created before the CRM migration; those are excluded downstream by* `account_type != 'internal_test'`. ***Expected range:*** *Between the first production account date and today.* ***Quality expectation:*** *99.5% of accounts populated; 0% with future dates.*

That's it. One field, one paragraph. A 50-field dataset takes a few hours. Most enterprises don't do it, and the substitutes all fail the same way. They *trust the data*, assuming field names are self-explanatory and everyone means the same thing by "customer segment." Wrong on at least one count. They *document verbally*, so the definition drifts through three retellings and dies when the person who knew it leaves. They *document in code comments* that aren't reviewed and become lies six months after the formula changes. Or they build a sprawling Data Governance Spreadsheet™ nobody updates, its last-modified date two years stale and half its rows describing systems that were decommissioned since. None of these are dictionaries. They're approximations that collapse exactly when you need them.

The reason to build one isn't pedagogical; it's operational, and it's four things at once. A dictionary is how you know what you're *allowed* to do with data: that "customer_segment" was created for marketing with consent implications, so nobody repurposes it for training and manufactures legal risk. It's how you know when your data is *wrong*, because without defined expectations you won't notice a field break. It's how you know what you can safely *combine*, because a field whose calculation changed six months ago means the pre- and post-change records represent different things, and training on both degrades the model quietly. And it's the only real *enforcement* mechanism: "all AI data must be documented" is unverifiable without the artifact that proves it.

A per-dataset dictionary doesn't fully solve the deeper problem, which surfaces the moment data crosses system boundaries: the same concept gets called three different things. "Customer age" is `customer_age` in the warehouse, `age` in the ML pipeline, and `customer_demographics.age` in the API, each locally sensible, collectively a trap, because it makes tracing lineage nearly impossible. The enterprises that get this right build a common semantic layer (one source of truth for what “customer age” means everywhere), so “where is customer age used?” has exactly one answer instead of an archaeology project.

None of this requires a massive program, just discipline at the point of use: when a dataset enters your pipeline, someone (a data engineer, data scientist, or governance specialist) owns documenting what each field means, where it came from, and what quality to expect, two to four hours per dataset. Unglamorous work, and the foundation everything else sits on. The enterprises that do it have simpler conversations: compliance asks “is this documented?” and the answer is yes; a model degrades and you can tell whether the model is wrong or the data changed.

From Tax to Lever

Everything so far reads like overhead: metadata, lineage, definitions, an owner for each dataset. Most enterprises treat it exactly that way, and the cost shows up the moment the ground shifts.

Consider a regulator asking for something genuinely new, a fairness requirement: *when you make decisions using AI, verify the model has no disparate impact on a specific demographic group.* If you understand where every field came from, what it means, how it’s used, and what it’s called in each system, you trace it in an afternoon (“our model uses field X, from source Y, mapping to group Z”), check fairness, and answer in weeks. If you don’t, the same question takes months: three systems call the same thing by different names, the demographic data lives in a database that’s hard to access, and the deadline passes while you’re still figuring out what you have. That gap (“we can respond to this” versus “we can’t move fast enough”) is the whole game.

What makes governance strategic rather than ornamental is specific: lineage tracked *with purpose*, so a fairness question maps instantly to every place the data is used; quality thresholds tied to *use case*, because quality isn’t absolute: customer age needs less precision for marketing than for fairness auditing, which lets you make deliberate trade-offs (“good enough for fraud detection, not for fairness auditing”) instead of across-the-board upgrades; access trails, so you can say who used which demographic data last month; and ownership with a name attached.

All of it is expensive, and it needs teams willing to move together on semantic consistency even when they’d rather move independently. So it gets deferred, because when a model breaks it's blamed on model design, and the bill doesn't come due until a regulatory change lands or an integration breaks or you need to trace an incident and can't. Then you discover the cost of building it in calm is a fraction of building it in crisis. Over three to five years the difference compounds, and the enterprises with strong data governance pull ahead — not because they're more compliant, though they are, but because they're more *adaptive*. You'll change your AI governance several times in the next five years; data governance is what turns each change from a from-scratch rebuild into an adjustment within a clear frame.

Start With the Data

Every significant governance failure in enterprise AI traces back to the same root: a foundational data problem identified too late. Not merely a quality issue: a deeper question about where the data came from, who's represented in it, and whether it reflects what the model is supposed to do. Ask those questions before a model ships and you're managing a system. Ask them after, and you're managing a crisis.

The hardest version is *inherited data*: data that's flowed through your organization for years, that you didn't create, that nobody fully understands anymore. A stream joined and transformed so many times the original source is lost, a dataset collected for one purpose and now used for another entirely. Inherited data is where most governance gaps live, and building AI on top of it is a structural risk no amount of model governance can fix. Leading organizations run actual data audits on the high-risk sets: going back through them, documenting where they came from, surfacing the baked-in assumptions, flagging which are dangerous downstream. It's expensive. It's also foundational.

Mature data governance, stripped to what matters, is five things you can say about any dataset used in consequential decisions. **Provenance:** where it came from, how it was collected, who created it, under what constraints. **Representativeness:** what population it represents, who's underrepresented, and therefore what a model built on it can and can't safely do. **Freshness and drift:** whether it still represents what it used to, monitored continuously rather than assumed once. **Lineage:** for derived data, the ability to trace back through every transformation and know what each one assumed and what the aggregation lost. **Access and accountability:** who's using which datasets for what, and who's responsible for the foundational ones. You don't do this for your whole warehouse; you start where the blast radius is largest and document those five things for each. That's the point where governance stops being theoretical and starts being operational.

Data governance is harder than model governance precisely because it requires understanding things that live outside your systems: how data was collected, what the collection missed, how the world has changed since. That difficulty is also why it's foundational. The organizations moving fast are moving fast *because* they understand their data: they know what their systems can safely do because they know what the data represents. All the rest (model governance, compliance, risk management) is built on top of that understanding. Start with the data. Everything else builds from there.

The Bottom Line

Enterprise AI governance fails at the data layer, not the model layer, because the model is only an amplifier: the real risk surface is the data underneath it. Two failures cause most of the damage: you can't prove where your data came from, and you can't agree on what it means. The fix isn't a better quality score or a bigger framework; it's the discipline of mapping defects to decisions, documenting what fields mean, versioning training data as an artifact, and giving each critical dataset a named owner. Treat the criteria for putting data into production as identical to the criteria for putting a model into production. Do that boring work before deployment and you're running a system that compounds value; skip it and you're one regulator's question away from a crisis you can't diagnose.

Field Kit

1. The five-question dataset intake. Bring these before a dataset feeds a model. (1) Where did this come from, and who was incentivized to enter it — to do what? (2) What changed in the source system in the last three years? (3) What does each field mean, and is it defined the same way everywhere it's used? (4) Can we produce a manifest of exactly what trained this model, and reproduce the historical processing? (5) What licensing and consent restrictions apply to each part? If any answer is "we'd have to go find out," the dataset isn't ready.

2. The defect-to-decision map. A two-column diagnostic. Left: known defects, tagged structural / semantic / temporal. Right: the decisions this system makes, tagged by cost-of-failure (low / medium / high). Draw the lines. Where a high-cost decision touches a semantic or temporal defect, scope that decision down — advisory not binding, one business unit not all, shadow mode extended — until the defect stops mattering.

3. The one-field dictionary template. For every field entering an AI pipeline: *field name · plain-English definition (the methodology, not the label) · data type · source table/column or formula · notes and known changes · expected range · quality expectation (percent populated, what should never happen).* Store it in version control; review it when data changes. Budget two to four hours per dataset, one named owner each.

4. The dataset owner's veto. You already name owners; at the data layer the thing to specify is what that name is empowered to stop. For each critical dataset, the owner's "this source isn't ready" is a hard deployment blocker with the same standing as a failed model validation — raised before training, not after launch. Test whether it's real the cheap way: ask when they last used it, and what happened next. An owner who has never blocked anything, or who was overruled without a written rationale, is a label rather than a control.

5. The high-risk triage list. You can't govern everything at once, so rank by blast radius: datasets used in consequential decisions (who is harmed if the model is wrong?), datasets that feed multiple downstream systems (cascade risk), and datasets you inherited and don't fully understand (structural risk). Work top-down; document provenance, representativeness, freshness, lineage, and access for each before moving on.

6. The regulator dry-run. Once a quarter, pick a live model and answer, cold and timed: *"You recommended X for this segment — what data trained the model for that segment, what does it represent, and why is the recommendation fair?"* How long it takes, and how much you guess, is your real governance maturity score. Weeks means you built the foundation. Months means you're building it in the next crisis instead.

PART IV

Vendors, Foundation Models & Independence

CHAPTER 9

The Foundation-Model Dependency Trap

Part IV · Vendors, Foundation Models & Independence

A CTO I know recently walked into a board meeting and said something the room wasn't ready for. "We've built our entire competitive advantage on a foundation model API. What happens when our vendor changes their pricing? What happens when they change their terms? What happens when they decide to compete with us directly?"

The board had no answers. Not because they were unprepared people — they were seasoned, careful, the kind of directors who ask good questions about currency exposure and customer concentration. They had no answers because nobody had ever modeled the scenario. It felt theoretical. Or distant. Or like someone else's problem.

It is not theoretical, it is not distant, and it is nobody else's problem. It is the central architecture decision facing most enterprises building on AI right now, and the majority of companies have not thought through the implications. They have priced the license. They have not priced the dependency.

That is the trap. And the first thing to understand about it is that it was not an accident.

Here is the reframe this chapter is built on.

The myth: choosing a foundation model is a procurement question about capability and price, and the dependency it creates is something you manage afterward.

The reality: vendor dependency is a governance decision, and the lock-in is structural, so what matters is whether someone priced that constraint and owns it or it happened by accident.

The Lock-In Is Structural, Not Accidental

Let's be precise about what has happened over the last couple of years. Foundation models have become genuinely useful, genuinely reliable, and genuinely easy to drop into a business system. A team of three engineers can now ship a production feature that was flatly impossible to build a few years ago. That is real progress, and you should use it.

But look at *why* you're using it, because the reason is also the reason you're stuck.

You're not treating a foundation model as one option among five interchangeable ones. You're building on top of it because the capability is native to the model. The few-shot reasoning, the instruction-following, the breadth of knowledge: you cannot replicate that behavior with a smaller model or a weekend fine-tune. It is a property of scale, and scale is exactly what you don't own.

Then your domain adaptations optimize for that specific model. You tune prompts to its strengths. You build workflows around its output format. You architect your system around its API constraints. Moving isn't a swap; it's a rearchitecture.

And then there's the part nobody puts on a slide: your team builds muscle memory. Your engineers think in that vendor's tokens. Your product managers know the API's quirks by heart. Your operations people have learned to live with its cost profile. There is switching cost buried in the code, and there is switching cost buried in your people.

Add these up and something quiet happens. "We can swap vendors if we need to" becomes "this vendor is our foundation." The word choice matters. You don't build a house so you can move it next year.

So the question that should keep you up at night isn't hypothetical, and it isn't paranoid. It's an accounting question: *what is your actual downside if the terms change?*

What "Terms Change" Actually Looks Like

If that still sounds abstract, it shouldn't, because every one of these has already happened to somebody.

A provider raised pricing on its flagship API, and companies that had modeled their unit economics around the old rate had to reprice their own products or eat the margin compression. A provider changed its rate limits, and production systems hit bottlenecks their builders never designed for. A provider changed its token-counting methodology, and the monthly bill moved under systems that had assumed it wouldn't. A provider released a cheaper model with similar capabilities, and companies that had built a moat on the assumption of exclusive access to high-quality reasoning suddenly found themselves competing on price instead. A provider re-priced its long-context features, and teams that had optimized hard for that specific cost structure had to rebuild around new math.

Here is the part people miss when they hear these stories: none of it is villainy. These are normal business decisions by vendors optimizing revenue and managing demand. That's the point. You cannot wait for your vendor to behave badly, because the damage doesn't require bad behavior. It only requires the vendor to act in its own interest, which it will, on a schedule you don't control.

And every one of those normal decisions forces an expensive, unplanned recalibration onto the companies downstream. Yours would be one of them.

But the scenario that should worry you most is the one that hasn't happened to you yet.

The Exposure Nobody Prices: When Your Vendor Becomes Your Competitor

Here is the asymmetry at the heart of the trap. You are optimizing your entire business on top of someone else's infrastructure. Your customer relationships, your domain expertise, your hard-won operational knowledge: all of it lives on their API.

And they can see it. They see your usage patterns. They see what you're building. They understand your market, because your traffic is teaching it to them. They have the infrastructure, and increasingly they have their own customer relationships.

So if your vendor decides your market is worth having, they can build a competing product faster and cheaper than any outside challenger, with native integration to the model, direct API access, no external dependency to manage, and a direct line to the customers you share. You will have market-tested the opportunity on their behalf, for free.

This is not a conspiracy theory. It is the ordinary trajectory of platform companies. It happened in cloud computing, where providers that started by selling foundational infrastructure moved up the stack into the applications their own customers had built on that infrastructure. It happened in search, where dominant platforms built ad networks and then began competing with the very services that lived on their search traffic. Platforms move up. That is what platforms do.

The cruel detail is that the switching costs run in the vendor's favor. Because you built a bespoke integration layer for one vendor's API, migrating to another isn't an engineering ticket — it's a rebuild. They know that. You should price that risk into your strategy before they price it into theirs.

The Three Comfortable Answers (And Why None of Them Hold)

When enterprises do finally confront this, I see three patterns. All three feel responsible. None of them actually solves the problem: they just hide it in different places.

The hedge. "We'll use multiple foundation models." Two vendors means two roadmaps, two sets of terms, and two independent incentives to move up the stack into your market: fragmentation wearing redundancy's clothes, which Chapter 11 takes apart in full.

The deferral. "We'll start with one model and re-evaluate in a year." This is rational caution, and it is exactly how you end up trapped. Year-one integrations become load-bearing. Dependencies calcify. By the time your calendar reminder to "re-evaluate" comes up, your systems have dependencies you've forgotten you have, and the evaluation you scheduled is now a migration you can't afford. Of the three answers, the deferral is the only one that gets strictly worse through nothing but the passage of time, which is why it is both the most common and the least noticed. Nobody ever holds a meeting to decide to become more locked in.

The escape hatch. "We'll train or fine-tune our own model." Intellectually satisfying, occasionally correct, and, as Chapter 10's arithmetic shows, almost always underestimated. But set the arithmetic aside for a moment, because there is a failure mode here that survives even when the numbers work out, and it is the version people are least prepared for.

I watched it play out with a team that took the escape hatch. They built a suite of applications on top of an open-source model. They contributed improvements upstream. They optimized hard for their use case. Six months in, they had twelve applications in production, every one of them depending on that model. Then the maintainers made a change — not a breaking change, exactly, but one that altered how the model handled certain inputs. The team faced a fork in the road, literally. Spend weeks updating all twelve applications, or fork the model and maintain their own version.

They forked. Their fork began to diverge. Community updates now had to be merged by hand. New bugs surfaced in the original model, and every time they had to decide whether to backport the fix. Two engineers ended up spending a meaningful share of their year as maintainers of software their company did not sell. Within two years, maintaining that fork had cost more than building their own model from scratch would have, for the structural reason Chapter 10 names precisely. But notice what nobody on that team was tracking through any of it. They hadn't been paying attention to the governance implications of the dependency. They'd been paying attention to whether the model worked. It did. That was never the question.

Which brings us to the reframe the whole chapter turns on.

This Is a Governance Problem, Not a Procurement One

Most enterprise AI vendor relationships are built backwards. A company has a gap. It needs a model. It surveys the options, picks the best one on the benchmark sheet, and *then* manages the dependency that choice created.

That is exactly the wrong order. The question was never "which model should we buy." The question is: "if we depend on this vendor, what constraints does that create, and can we live with them?"

Say it plainly, because it changes who needs to be in the room: vendor dependency is a governance decision, not a procurement decision. When you choose a vendor's model, you are not simply choosing a level of performance. You are choosing to accept that this vendor can change how your model behaves, and you'll have to keep up. You are accepting that if their business changes, yours might have to. You are accepting that you now live on someone else's roadmap.

None of that is disqualifying. Dependencies are inevitable; you should be dependent on foundation models, because they are better at many tasks than anything you can realistically build or buy. The problem is never the dependency. The problem is the *accidental* dependency: the one nobody decided, nobody priced, and nobody made anyone accountable for.

The Questions You're Probably Not Asking

Here's a scene I've watched more than once. You're across the table from a vendor. You ask about the composition of the model's training data (in detail, not "internet-scale data"). You ask how they audit for performance gaps across demographic segments. You ask what happens, and what they owe you, if the model's outputs degrade in production.

Three questions, three versions of the same answer: "We haven't really thought about that."

They are not bad actors when they say it. Most vendor relationships simply haven't matured to the point where these conversations are routine. But the blank stare is itself data, and you should treat it that way.

I sat in on an evaluation where an organization was choosing a model for a mission-critical, regulated use case. Three vendors, all with impressive benchmark scores, two with meaningful market share. All three came apart on the governance questions. Vendor A couldn't explain how they'd support model evaluation in a regulated industry. Vendor B had no framework for even knowing when their model was performing poorly for a specific kind of decision. Vendor C had no clear terms for what happens if the model causes harm in production: no answer on insurance, on liability caps, on shared responsibility. Not villains. Just vendors who had built good models and had never been asked to integrate them into someone else's risk framework.

The reason this matters so much is an imbalance you cannot negotiate away: *you don't get to choose who builds the foundation model, but you carry full responsibility for how it's used.* Your organization is accountable. Your model's outputs are driving real decisions. Your brand is on the outcome. If the model fails, you are the one explaining it to a regulator, a customer, and your board — not the vendor. Yet you cannot make the vendor invest in understanding failure modes, in bias auditing, in degradation monitoring, in liability. You can only decide whether to use their model with your eyes open or shut.

So open them. Benchmarks are table stakes: the end of your technical evaluation, not the beginning. The real evaluation is about robustness, transparency, and what happens when things move. These questions aren't just for the frontier-API vendors, either; ask them of open-source models, cloud services, and your own internal platforms alike. No provider is exempt.

Ask about transparency first. What data was the model trained on — the composition, what's in and what's out, when it was trained, what's changed since? You cannot assess a model's risk if you don't know what shaped it.

Ask about bias. Not "do you think bias matters." Everyone thinks bias matters. Ask for the operational framework: how do they detect performance gaps across populations, do they have the data, will they share it, what's the protocol when a gap shows up? Many won't have a crisp answer. That tells you something.

Ask about degradation. How would they know if the model's outputs started getting worse, how fast, and what do they owe you when it happens? Some monitor internally but won't share. Some don't monitor your deployment at all: that's your job, they'll say. Find the boundary before you're standing on the wrong side of it.

Ask about accountability. If the model causes harm in production, what is theirs to own? Insurance? Liability caps? Shared responsibility? This is the question that feels adversarial, which is exactly why it's the one to bring your general counsel to. If the vendor hasn't thought about it, that's information. If they have crisp answers, it means they've dealt with enterprises who needed them.

Ask about change and continuity. How often does the model change, and what counts as a change (weights, architecture, input and output handling)? What's the cadence: quarterly, on a whim, only on major improvements? Which changes are backwards compatible, and which will force you to retest and redeploy every dependent application? And the one most vendors will squirm at: what's your path if they deprecate this model or change their business model out from under you? How much notice do you get, can you download and run the weights yourself, what are the licensing limits? If the honest answer is "you'd have to rewrite everything," that belongs in your decision rather than in a footnote.

Ask about your data. If you're on their API, are they storing your inputs, for how long, and for what? If they're training on your data, that can be fine, but you should know it and be comfortable with it, not discover it later.

Ask about your specific failure mode. You know your business better than they know it. Describe the exact place this model could hurt you, and ask whether they've seen it, what they'd recommend, and how they'd support you if it happened. Some will have thought it through. Others will hand you generic reassurance. Both answers are useful.

You are not looking for perfect answers here; you won't get them, and that's not disqualifying. You're looking for evidence that this vendor has met enterprise risk management before, because when you select a foundation model, you are not just selecting a model. You are selecting a risk partner. And the good news, which surprises people, is that when you press clearly on what you need, most vendors want to get better at this. They built strong models and never had to wire them into anyone's governance. Ask precisely, and many will work with you on transparency, monitoring, and accountability. The ones who engage tend to be the ones you keep.

What Designing for Portability Actually Requires

None of this is a reason to build less on foundation models. It's a reason to build so that you can still change your mind.

The mechanism is an abstraction layer you own, sitting between your applications and the vendor. Chapter 10 is where that gets built: what it spans, what it costs to carry, how often you should exercise it. What belongs here, in a governance chapter, is the part that never arrives on an engineering backlog by itself: portability is a property you *buy*. It has a price, it appears on no feature roadmap, and it is therefore the first thing traded away when a quarter gets tight. The real work of dependency management is not visionary. It is a budget line with a name beside it.

Which makes portability a board-visible choice, and it has to be somebody's job. I have not yet seen an organization hold onto real portability without a named executive who could stand in a budget review and say, "this line is what keeps our options open, and here is what it costs us." Where nobody owns that line, the pattern is depressingly consistent. The abstraction gets designed. It gets half-built. Then a launch date moves, someone observes correctly that calling the vendor's API directly ships two weeks faster, and the shortcut is taken on the explicit understanding that it will be cleaned up next quarter. It is never cleaned up. Two years on, the "temporary" direct integration is in nine services, and the abstraction layer exists only in an architecture diagram nobody has opened since.

The number that wins this argument with a CFO is the switching premium. If you can only migrate at a 40% premium, then 40% above today's API price is your real cost baseline, not the figure on the invoice. Carry *that* number in your unit economics and portability stops being an engineering preference and becomes a line item with a return attached, which is the only form in which it reliably survives budget season.

Alongside it, the organizations that stay out of trouble do four things at the governance level. They **map their dependencies**: they know exactly which applications lean on which models and what breaks if those models change. They **diversify where it's sensible**: where an application could genuinely run on more than one model, they build for that flexibility instead of pretending the option exists when it doesn't. They **keep real fallback options**: a specific plan, whether "migrate to this alternative," "stand up our own version," or "redesign the application," not a vague intention. And they **make dependency an explicit choice**: every place they accept lock-in, someone decided to accept it, on the record, knowing the constraint.

If you're reading this already deep in a dependency (twelve applications on one model, terms you never negotiated), you can't undo it quickly. But you can start now: map the dependencies, build flexibility in where it's cheap to add, ask the hard questions you skipped the first time, and convert the accidental dependencies into deliberate ones, one at a time. The companies that end up in bad situations almost never chose them. They just picked a model that worked and kept building, and never looked up.

The Conversation You Should Be Having Now

So the question is not "should we depend on a foundation model?" You should. The question is: "what is our dependency-management strategy?"

That is not, at its core, a technical conversation, which is why leaving it to the engineering team is how it never happens. It's a strategy conversation, and it needs four people who rarely sit down together for it. Your CTO owns the architecture and the abstraction layer. Your CFO owns the total cost, including the switching premium that isn't on any invoice yet. Your product head owns what a vendor's change does to customers. And your board owns the risk: the one the CTO named at the top of this chapter, that nobody had modeled.

Have that conversation before the terms change, not after, because after is just a post-mortem. The companies that compound real value from AI over the next several years won't be the ones who built the fastest integration this year. They'll be the ones who built it in a way they can still change next year.

The Bottom Line

Depending on a foundation model is not the mistake. It's unavoidable and often correct. The mistake is depending *accidentally*: accepting a strategic constraint no one priced, no one decided, and no one owns. Vendor lock-in is structural, not incidental, and it deepens while your systems, your economics, and your team all optimize around a platform whose owner answers to its own incentives and can, at any point, raise your costs, break your assumptions, or move up the stack into your market. Treat model selection as a governance decision, not a procurement one: interrogate the vendor as a risk partner, fund portability as a named budget line with a named owner instead of a good intention, and put the switching cost into your unit economics as a first-class number. The goal is not independence, which you can't afford, but *optionality* — the ability to still change your mind after you've committed.

Field Kit

1. The board question set. Bring these four to your next board or strategy review; if no one can answer them, that gap is the finding: - What happens to our economics if our primary model provider raises prices or changes billing? - What happens to our product if they change the model's behavior, deprecate the version we use, or alter our terms? - What happens to our market position if they build a product that competes with ours? - If this model causes harm in production tomorrow, which part of that is ours to own and which is the vendor's — and where is that written down?

2. The vendor risk-partner interview. Use in every model evaluation, after benchmarks, not instead of them. Score the *quality of engagement*, not just the answer: - **Transparency:** training-data composition, cutoff, and what's changed since. - **Bias:** the operational framework for detecting and addressing performance gaps across populations — with data. - **Degradation:** how they'd detect declining outputs, how fast, and what they'll do; where the monitoring boundary sits between you. - **Accountability:** liability, insurance, caps, shared responsibility if the model causes harm (bring your general counsel). - **Change & continuity:** cadence of change, what counts as a change, backwards-compatibility, and your exit path if they deprecate or re-price. - **Your data:** what's stored, for how long, whether it trains their models. - **Your failure mode:** describe the exact way this model could hurt your business and ask how they'd support you.

3. The dependency map. A one-page inventory: every production application, the model(s) it depends on, the blast radius if that model changes, and whether a real fallback exists or only a hope. Refresh it quarterly. If you can't fill it in, you don't yet know your own exposure.

4. The real-cost worksheet. Take today's API price and add the amortized switching cost. If migration would run a 40% premium, your true baseline is 1.4x the invoice — carry *that* number in your unit economics and your board deck, not the sticker price.

5. The portability line item. Portability is bought, not wished for. Name the executive who owns it, put its cost in the budget where the CFO can see it, and require a written sign-off — not a verbal one — whenever a team ships a direct-to-vendor integration to make a date. Chapter 10 supplies the engineering checklist that owner should be holding teams to; your job is to make sure the line survives the quarter in which it is inconvenient.

6. The accidental-to-deliberate audit. For each vendor dependency, name the person who decided to accept it and the constraint they knowingly accepted. Any dependency without a name attached is an accidental one — convert it: diversify it, build a fallback for it, or consciously sign off on it.

CHAPTER 10

Architectural & Strategic Independence

Part IV · Vendors, Foundation Models & Independence

A CTO once told me they were building their own fine-tuned version of an open-source model to reduce vendor dependency.

I asked why.

"So we're not locked into one vendor."

That is the right concern. The execution is backwards.

Building your own fine-tuned model does not make you independent. It makes you dependent on a different thing — your ability to maintain it — and that is a dependency most enterprises underestimate. You have not escaped a trap. You have traded a vendor you can renegotiate with for an obligation you can never put down.

Here is the reframe this chapter is built on.

The myth: independence from AI vendors is a procurement decision, won by picking the right vendor or by building your own model instead of buying one.

The reality: independence is a property of your architecture, not your contract, and it is decided by how you wire a model into everything else you own.

This is the pattern that runs through almost every conversation about AI independence, and it is almost always miscast. Leaders treat independence as a procurement question (which model, which vendor, build or buy) when it is really an architectural one. The decision that determines whether you are free or trapped in five years is not which foundation model you pick. It is how you wire that model into everything else you own. Independence is a property of your architecture, not a line item in your vendor contract. You can build your own model and still be hopelessly locked in. You can run entirely on a proprietary vendor and stay free to leave. The difference is not the model. It is the design.

The Symptom and the Disease

Most organizations think independence means: don't use foundation models, build or fine-tune your own. That is treating the symptom, not the disease.

The disease is structural dependence on any single infrastructure or capability that you cannot replicate if the vendor relationship changes. The cure is designing your systems so that your core capabilities either exist across your organization as institutional knowledge, can be replaced without rebuilding the entire application, or are decoupled from your core business logic.

Building your own model can accomplish that. But so can building abstraction layers. So can designing for model rotation. So can maintaining real technical depth in how your models actually work. The question is not "should we build or buy." The question is "what do we need to maintain control over, and how do we actually maintain it?" Once you frame it that way, building your own model stops looking like the answer and starts looking like one very expensive option among several.

What Building Your Own Actually Costs

Be concrete for a moment about what it means to build and maintain your own fine-tuned models, because the brochure version and the reality diverge fast.

The initial build is the smallest of it. You need engineers who understand model training: data scientists with five or more years of experience, the kind you are already competing for. If you are building from scratch rather than fine-tuning, you are looking at $1–3M in development cost, six to twelve months of timeline, and a team of four to eight people. That is the part everyone budgets for.

Then the costs that don't show up in the proposal begin. A foundation model API bills you on usage; running your own bills you on compute: GPU infrastructure at roughly $100K–500K annually depending on scale, plus the electricity, networking, and datacenter overhead that ride along with it. Then the maintenance burden, which never ends: when a better model ships, do you retrain? When your domain shifts and your training data goes stale, do you retrain? When you find bias in your model, who fixes it and how? Each of those is a live project with an owner, forever.

Then talent concentration, which is the quiet killer. You now hold a capability that only a handful of people in your organization understand deeply. If your model expert leaves, gets overloaded, or moves to another team, you have a problem you cannot buy your way out of quickly. You have traded vendor dependency for person dependency, which is worse, because you cannot renegotiate with a resignation letter.

And finally opportunity cost. Every month your best engineers spend maintaining model infrastructure is a month they are not building the applications that infrastructure exists to serve.

The argument for building your own always sounds like "this gives us independence." The reality usually reads: this trades one dependency for another, often a more expensive one, and concentrates the risk in a place that is harder to see and harder to exit.

When Building Your Own Is the Right Call

There are real cases for building your own. But notice, when you line them up, that not one of them is actually about independence.

The first is **proprietary advantage through data.** You hold data nobody else can access, and that data is a competitive moat. A financial services firm with fifty years of internal transaction history. A manufacturer with sensor data unique to their own equipment and process. Here you are not using a foundation model as a convenience; you are folding in proprietary data to create a capability competitors cannot replicate, and the privacy and protection are worth the build cost. But even here, you can usually take a foundation model as the base and fine-tune on your proprietary data. You rarely need to train from scratch.

The second is **extreme scale or cost sensitivity.** You run the model so many times that the marginal cost of API calls dwarfs the cost of running your own infrastructure: you have done the math, and your compute comes out 30% cheaper in-house. A company processing millions of requests a day, where every call carries a price. That is real economics. But it still buys you the operational burden of maintaining the infrastructure, whether you wanted that job or not.

The third is **regulatory or data residency requirements.** Your industry demands that models be trained on data that never leaves your environment. A healthcare provider with patient data, a financial institution with transaction records, a government agency with classified material. This reason is absolute — you have no choice. But even here you can often start from a self-hosted open-source base and add your domain-specific fine-tuning on your own infrastructure.

In all three cases the justification is data: proprietary, sensitive, or scale-driven. None of them is "independence from vendors." If independence is your goal, building your own is almost never the shortest path to it.

The Decision Nobody Realizes They're Making

Chapter 9 described the dependency; this is the decision that creates it.

So where does the real trap live? Not in the build-or-buy debate everyone argues about. It lives in a decision that looks small at the time, gets made by a technical team, is noted in a meeting, maybe documented somewhere, and is never revisited. The decision is: *where do we put this model, and does it talk to our applications directly or through something we own?*

That decision will cost you millions, and most organizations make it wrong without ever realizing it was a decision at all.

It happens through compounding convenience. You adopt a vendor's platform because they have the best model for your use case (a smart call). You wire your data into their infrastructure, because that is where the model lives. You build a second model on the same platform, because you are already there and moving the first would be chaos. By month eighteen you have seventeen models on that vendor's infrastructure, your data team knows their APIs, your DevOps team understands their monitoring, and your compliance team understands their audit trail. You have solved every hard problem — using their tools. The cost of leaving is no longer the cost of migrating one model; it is the cost of rebuilding your entire AI operating system on new infrastructure.

Notice what actually did the welding, though, because it was not the contract. The contract runs a year and you can renegotiate it. What welds you is the count of places in your own estate that name a specific vendor: every application importing their SDK, every pipeline that expects their file format, every dashboard built on their metric names, every runbook written around their failure modes, every engineer whose only debugging instinct is their console. That count is a number *you* control, and almost nobody measures it. Two companies can run identical workloads on the identical vendor and sit in completely different positions: one has forty services calling the vendor directly, the other has one service calling the vendor and thirty-nine calling that service. Same contract, same spend, same model. Wildly different freedom. Lock-in is not something done to you from outside; it is a number inside your own codebase, and left alone it only goes up.

The Abstraction Layer

Reduced to its essentials, architectural independence is one modest idea: **you can change your mind about any single layer of your AI stack without forcing massive rewrites everywhere else.** It does not mean running multiple vendors for everything; that just multiplies your integration and training burden. It means no single layer is welded to the ones around it.

Concretely, the decision that matters is whether you build an abstraction layer between your applications and the model serving infrastructure. If you do, switching vendors costs about as much as switching databases (annoying but doable). Your applications talk to your abstraction layer; your abstraction layer talks to any inference infrastructure underneath; you swap what is behind it without touching application code. If you don't, your applications talk directly to the vendor's serving layer, and switching means every application team relearns new APIs, new authentication, new failure modes.

That independence shows up across four layers, and each follows the same logic. For **model serving**, your applications call your own serving abstraction, and the abstraction calls the vendor; switch vendors by changing what sits behind the abstraction, not what your applications call. For **training infrastructure**, you can use a vendor's platform to fine-tune, but you keep your model weights in a portable format so you can move them to another vendor or to open-source infrastructure. For **data**, you feed the vendor's platform but keep your own copy of the truth, so you are never hostage to their export capabilities, retention policies, or egress pricing. For your most critical **models**, you maintain the ability to run on more than one inference platform, which may mean favoring open-source, or keeping versions tuned for different backends. That is redundancy, and redundancy is the point.

Here is the honest part: this is boring work, and most teams skip it precisely because it is invisible. That is the wrong calculation. The abstraction layer never gets a moment where it looks impressive; it just means that five years from now, "we should move" is a decision you get to make instead of a wish you get to have. It is also cheaper than it sounds. A thin serving layer in front of an open-source runtime (vLLM, Ray Serve, BentoML, TensorFlow Serving) is a few weeks of work, not a platform program. Use the vendor's platform for fine-tuning, for training, for data preparation. Just do not let it be the thing your applications call.

The Open-Source Mirage

There is a comforting story that says open-source solves all of this for you. Run the model anywhere, own the compute, no per-call fees, no dependence on a vendor's API. The logic is sound in theory. In practice, enterprises that adopt open-source to escape lock-in routinely discover they have built a *different* kind of lock-in: one that is harder to escape than the contractual kind, because it is architectural.

The confusion is this: open-source is a licensing choice, and lock-in is an architecture problem. A licensing choice cannot solve an architecture problem. Watch how most enterprises actually use an open-source model. They download it, fine-tune it on their data, integrate it deep into their applications, and optimize the whole pipeline around its specific behavior. The fine-tuning is tied to that model's architecture. The integration is built around that model's exact output format. The batch sizes, inference settings, and caching are all tuned to that model's performance characteristics.

Now try to switch to a newer, better model. You have to retrain, because your fine-tuning process is bound to the old architecture. You have to reintegrate, because the new output format and inference API are different. You have to re-optimize everything. That is precisely the switching cost you would have faced with a proprietary vendor. The open-source license prevented none of it. Your architecture created all of it.

The enterprises that actually avoid this do the same thing whether their models are open or proprietary: they separate the model layer from the business logic. The application never knows which model it is calling. It calls a prediction service that says "here is my input, give me a prediction," and behind that service you swap models without the application noticing. That takes a prediction service that decouples the app from any specific model, standardized input and output formats so applications depend on a normalized contract rather than a model's raw output, version management that treats the model as a deployable, roll-back-able artifact rather than a permanent fixture, and monitoring pinned to business outcomes rather than model-specific internals. With that architecture, licensing becomes almost irrelevant: you can swap open, proprietary, or a different vendor entirely. Without it, licensing is entirely irrelevant, because you are locked in regardless of who owns the code.

So the real insight cuts against the marketing. Open-source models are genuinely valuable, especially where you need to avoid depending on a vendor's infrastructure. But the licensing does not give you flexibility — the architecture does. Build modular, decoupled systems, monitor at the business level, and plan for switching from day one, and you will have flexibility with open-source or proprietary models alike. Skip that discipline, and you will be locked in with either.

The Math of Independence

None of this is free, and pretending otherwise is how you lose the argument with your CFO. Independence has a real, ongoing cost. You are not using the vendor's native optimization or their deep integration; you are solving some problems twice. Expect model serving to run perhaps 15% slower without the vendor's tuned infrastructure. That is a latency penalty, not a budget line, and it is worth keeping the two straight. The budget line is roughly 10–20% additional engineering overhead and 10–15% additional infrastructure cost to carry the abstraction, the portability, the extra copies and pipelines. Blended, call it **10–20% ongoing**, and then use that one number everywhere, in the architecture review and in the board deck alike. The moment your CFO hears three different figures for the same thing, the conversation stops being about architecture and starts being about whether you know what you are talking about.

That cost is real. Now weigh it against the alternative. If you become dependent and then need to move, the switching cost is not incremental: it is measured in eight figures, sometimes nine, in multi-year migration projects, or in staying on a legacy architecture forever because leaving costs too much. During any such transition, which drags on for a year or more, you cannot confidently update your AI infrastructure or adopt new techniques, because you do not know whether they will work in both the old and new systems. You operate in a carefully managed semi-paralysis while competitors who made different architecture decisions move faster.

So the trade is tens of millions in overhead against hundreds of millions in switching cost, and it turns on one probability: how likely are you to need to switch? If there is even a 20% chance you will need to change vendors in the next five years, independence is the right bet. And most large organizations sit far above that line: realistically 60–70%, once you account for pricing changes, feature gaps, competitive moves, acquisitions, reorganizations, and regulatory shifts. At those odds, the question answers itself.

What Actually Works

Strip away the model debate and the open-source debate, and strategic independence, for any organization not in one of the three data-driven cases for building its own, comes down to five practices.

Design model-agnostic abstraction layers. Write code against an abstraction, not a specific vendor's API, so different models can satisfy the same contract. It is more complex and slightly less efficient, and it is the single most practical source of real independence for most organizations.

Maintain real technical depth. Don't treat models as black boxes. Keep people who can debug models, understand prompt behavior, and interpret outputs. If your people truly understand how models work, you are not dependent on any one vendor, because you have the skill to work with alternatives.

Build model rotation into the roadmap. Treat switching as a routine architectural activity, not a crisis. Every eighteen to twenty-four months, evaluate the field, test alternatives, and understand your switching costs, so that when you do move you are not starting from zero.

Keep the model separate from application logic. The model is a pluggable component, not the core. Your business logic should be independent of which foundation model it runs on today or tomorrow.

Maintain financial optionality. Set aside 5–10% of your AI budget to evaluate alternatives, and make model choice a decision you revisit rather than a habit you inherit. If your only model is the one you happened to start with, that is dependency. If your primary model is chosen and re-evaluated on purpose, that is strategy.

Most boards never think about any of this, because architecture reads as a technical detail. But architecture creates lock-in, lock-in creates cost, and cost constrains strategy. When your AI spend comes in higher than expected in five years and someone asks why you can't just move to a cheaper serving platform, the honest answer will be "because of a decision we made in 2026 and never revisited." The organizations that get this right will not necessarily switch models at all. They will simply have the ability to — which means they negotiate better, keep strategic clarity, and adapt faster when the ground underneath them shifts.

If You're Already Locked In

Much of this reads as advice for greenfield programs, but most enterprises are not starting fresh: they already have their seventeen models on one vendor's infrastructure. The move there is not a heroic migration. Do not try to move the old models; the cost rarely justifies it. Instead, draw a line: from today, new models go into an abstraction layer first, new data pipelines use portable formats, new training keeps weights exportable. You accept the legacy estate as it is and refuse to add to it. In five years you are a hybrid. In ten, you can actually switch. You do not undo the trap in one motion — you simply stop digging, and start building the exit one new system at a time.

FIGURE 10.1

The Abstraction Layer

The same applications, either side of one architectural choice.

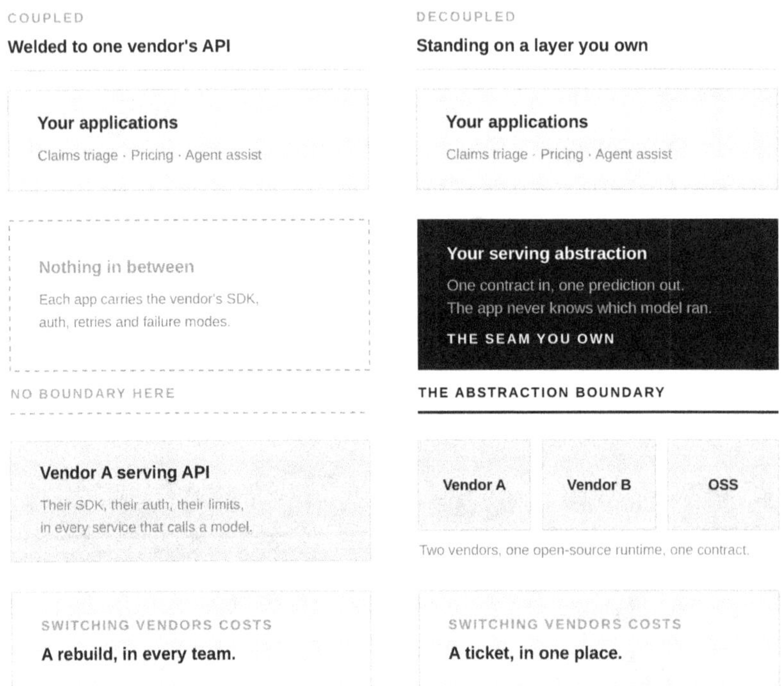

The layer is boring, cheap, and the whole point.

Carrying the abstraction runs 10–20% a year in engineering and infra.
The rebuild it prevents runs to eight or nine figures, on the vendor's timing.

Workshop use: engineers mark which layer each live integration actually calls.
Anything landing on the vendor box is a portability item with a named owner.

FIGURE 10.1

The Bottom Line

Independence in AI is not a vendor you choose or a model you build — it is a property you design into your architecture, one decision at a time. The trap is never a single bad choice; it is a compounding series of individually rational conveniences that weld your applications to one vendor's serving layer until leaving costs nine figures. Building your own model does not buy you out of it, and open-source licensing does not either: only abstraction layers, portable data and weights, business-level monitoring, and a genuine plan to switch actually do. That discipline costs 10–20% in ongoing overhead, against switching costs measured in the hundreds of millions and odds of needing to switch that sit at 60–70% for most large enterprises. The organizations that win this will rarely change vendors; they will simply keep the power to, and negotiate, adapt, and sleep better because of it.

Field Kit

1. The five questions to bring to your next AI architecture review. - Do our applications call the vendor's serving API directly, or our own abstraction layer? - Have we ever actually run a production model on a second inference platform, or is portability a claim we have never tested? - Where does the single copy of truth for our training data live: our infrastructure, or the vendor's? - Could we re-run our fine-tuning process on a different model, or is it welded to this one's architecture? - When did we last evaluate an alternative model, and when is the next evaluation scheduled?

2. The lock-in diagnostic (score each 0–2; total under 6 means you are accumulating hidden switching cost). - Model serving goes through an abstraction we own — not the vendor's native APIs. - We hold model weights in a portable format and export them on a schedule. - We keep our own copy of all training and feature data. - Monitoring tracks business outcomes, not model-specific internals. - Model choice is re-evaluated on a fixed cadence, not inherited by default.

3. The build-your-own gate. Before approving any "build our own model for independence" project, confirm it clears one of the only three real justifications: proprietary data as a competitive moat, extreme scale where in-house compute is genuinely cheaper, or regulatory/residency requirements that forbid external APIs. If the reason is "independence from vendors," reject it — an abstraction layer delivers that for a fraction of the cost.

4. The switching-cost math, on one page. Estimate three numbers and decide: annual cost of independence (target 10–20% engineering overhead plus 10–15% infrastructure), estimated switching cost if you had to move today (typically eight to nine figures at scale), and honest probability of needing to switch within five years (most large enterprises: 60–70%). If probability × switching cost exceeds the cost of independence, you build the abstraction. It almost always does.

5. The already-locked-in playbook. Do not migrate the legacy estate. Instead: freeze net-new dependency (every new model goes behind an abstraction layer, every new pipeline uses portable formats, every new training run keeps weights exportable), name an owner for the hybrid roadmap, and set a five-year checkpoint to reassess whether the old estate is finally cheap enough to move.

6. The board-ready one-liner. "We are not switching vendors — we are buying the option to. It costs us 10–20% in ongoing overhead and protects us from a switching cost we have a two-in-three chance of facing within five years."

CHAPTER 11

Vendor Strategy Without the Trap

Part IV · Vendors, Foundation Models & Independence

A manufacturing company brought me in to answer what sounded like a simple question: why couldn't they move off their AI platform, even though the economics no longer made sense?

They had started with the vendor five years earlier, when the alternative landscape was narrower and the choice was obvious. Now the alternatives were cheaper and faster. On paper, the case to switch was overwhelming. So they ran the transition analysis. It came back at $8M and 18 months.

They stayed.

This happens far more often than enterprises admit, and it is almost never because the new vendor is actually worse. It is because switching has hard costs that do not show up in the ROI calculator until you are twelve months into the migration and discovering them one painful surprise at a time. The company that cannot leave a vendor it has outgrown did not make a bad procurement decision five years ago. It made a series of individually reasonable ones, and never watched the number Chapter 10 called the real measure of lock-in (the count of places in your own estate that name one vendor) climb, one sensible decision at a time.

That is the real subject of vendor strategy, and almost everyone gets the framing wrong. Leaders treat it as a contract question: which vendor, what price, how many of them. The contract is the least of it. A real vendor strategy manages two things the contract barely touches: how expensive the relationship is to exit, and how mature it is as a partnership. Miss either and you end up exactly where the manufacturer did: technically free to leave, practically trapped, paying more every year for the privilege of not having to think about it.

Here is the reframe this chapter is built on.

The myth: vendor strategy is a contract question: which vendor, what price, and how many of them.

The reality: vendor strategy is the deliberate management of two things the contract barely touches: how expensive the relationship is to exit, and how mature it is as a partnership.

The Switching Cost You Cannot See

The standard analysis of vendor switching looks straightforward. New license costs. Data migration (moving existing datasets, models, and configurations). Retraining staff on new interfaces. Time to run both systems in parallel during the cutover. A responsible team goes further and budgets for regression testing, performance benchmarking, and the integration work to connect the new platform to everything downstream.

These are all real costs, and most enterprises estimate them conservatively. Then they start the transition and discover the costs that never appear on a spreadsheet: the ones that turn a weeks-or-months estimate into a bill three times larger. The gap is not a calculation error. It is that switching cost has hidden layers that only surface once you commit.

The first is architectural dependency. Before the scoping work started, the manufacturer's platform lead told me the vendor was not deeply integrated: it was "just the model platform." The inventory came back with 47 downstream systems making quiet assumptions about how that platform behaved: how it handled errors, how data flowed, what it returned when it returned nothing. The new vendor's API was functionally equivalent and architecturally different, which meant re-abstracting two layers of code nobody remembered were load-bearing. That line item was the largest in the $8M.

The second is data format translation. Your old vendor represented model outputs, uncertainty, and feature importance in a particular way. The new one represents them differently. Every system consuming that output was built assuming the old format, so updating them means re-testing each one, because slight format changes have disproportionate effects downstream.

The third is performance characteristics. The old system was slow but predictable. The new one is fast but bursty. Your rate-limiting, queuing, and caching logic was tuned for the old profile, and the new vendor hits your bottlenecks in places the old one never did.

The fourth is failure modes. The systems around your AI infrastructure were built to handle the specific ways Vendor A fails. Vendor B fails differently, and you discover exactly how six months into production, when something unexpected happens and nothing catches it.

The fifth is the quietest and the most expensive: organizational knowledge. At the manufacturer, one of the original engineers had worked out years earlier that the platform degraded badly above a certain request batch size, and had built a pre-processing step to chunk requests below the threshold. It was undocumented and nowhere on the architecture diagram. It lived in his head and in forty lines of uncommented code, and it had protected the throughput of a production line for four years. When you switch, all of that walks out the door with the people holding it. The new team relearns the new platform's quirks from scratch, and often rebuilds the old workarounds anyway, because keeping a mystery safeguard feels cheaper than finding out the hard way what it was for.

What makes these costs so easy to underestimate is not their size but their shape. Some are concentrated at the decision point: data migration happens once. But most are spread thin across months of steady work. If someone asks "how long will the migration take?" and the answer is "one engineer for two weeks," that is easy to budget. If the honest answer is "various engineers will spend 30% of their time on vendor-specific integration for nine months," that is much harder to quantify and much easier to wave away. And the whole time you are transitioning, the old vendor keeps evolving (other teams still use it, it still ships changes), so the longer the switch takes, the wider the gap grows between what you are leaving and what you are moving to. Some enterprises rebuild their transition logic two or three times because the thing they are migrating away from moved underneath them.

This is also why padding the estimate never closes the gap: a 30% contingency hedges the work you can see, not the layers you cannot, and it is those layers that take the total to three times your first number.

When you sit across from a new vendor, none of this is what they will tell you. They will say switching costs are low and show you case studies of companies that moved in ninety days. Those stories are not lies, but the companies in them were usually moving off something worse, running simpler architectures, or sitting on spare engineering capacity you do not have. Their experience is real. It is just not predictive of yours. Your switching cost depends on how integrated your current vendor is, how much of your estate depends on its specific formats and error handling and latency, and how long you can tolerate paying for two systems at once.

The enterprises that get this right share one habit: they estimate switching cost during vendor *selection*, not during the transition. Before they sign, they ask a single uncomfortable question: *if we needed to move off this vendor in three years, how hard would it be?* If the answer is "very," they negotiate better terms up front or choose differently. The point is not that you should never switch. It is that you should know the true cost before you are committed, and be honest about whether the new vendor's advantages actually clear it. Sometimes they do. Sometimes the smarter move is to optimize within the constraints of the vendor you already have. You cannot tell which until you price the exit — and the deeper the integration, the higher that price.

The Diversity Myth

Once leaders feel the shape of that trap, the reflex is almost automatic. *We should diversify our vendor stack.* It sounds like exactly the right lesson: instead of leaning on one vendor, use several. Spread the risk. Reduce lock-in. Keep your options open.

In practice, it usually does the opposite. It increases complexity without delivering any of the flexibility it promised.

Follow the logic to where it actually leads. You use Vendor A for training, because it is genuinely excellent at training, and your team learns the platform. Then you need serving, so you bring in Vendor B, which has clean APIs, and your application teams integrate against it. Now you have two systems that do not really talk to each other. Data flows from Vendor A, through your systems, into Vendor B. You are paying integration costs, managing dependencies between two vendors, and staffing people who have to understand two platforms instead of one.

You have not reduced lock-in. You have doubled your surface area. And the moment something breaks (one vendor changes its pricing, or its strategy, or its capabilities), you are just as stuck as before, because Vendor A's training and Vendor B's serving were never designed to be recombined. Moving is exactly as hard as it always was. What you built was not vendor *flexibility*. It was vendor *fragmentation*.

The underlying instinct is correct: depending on a single vendor is a real risk. But the conclusion drawn from it (use more vendors) is wrong. The real vendor problem was never about the *number* of vendors. It is about *abstraction*. If your systems depend on a vendor's specific APIs, formats, and infrastructure choices, you are locked in and you have created technical debt. That is true whether you use one vendor or five. If there is a clean abstraction between your systems and the vendor, you can change vendors with reasonable effort (again, whether you use one vendor or five).

Most organizations try to solve a lock-in problem through multiplication. The answer is architectural. You do not escape dependence by adding vendors; you escape it by separating your systems from vendor specifics. And that holds no matter which vendor sits at the center of your stack.

What Actually Works

The right answer is less exciting than a multi-vendor slide, and it starts with a decision most procurement processes never make explicitly.

Choose one primary vendor for each critical function. Do not try to be vendor-agnostic across the board. Pick one vendor for model training, one for serving, one for data pipelines; for each, pick the one that is genuinely best-in-class for what you actually need. This is where you concentrate your expertise and your integration work. Using one vendor well beats using five vendors mediocrely, every time.

Put an abstraction layer you own between your systems and that vendor, built as Chapter 10 describes. Everything below rests on it.

For each critical function, name the alternatives that could actually replace your primary vendor. Not imagined alternatives. Specific ones. Could you move to Vendor X? What would it take? How long? What would be genuinely hard? For most functions you will find that most vendors are not realistic alternatives at all, and that is fine. You only need to know, for the handful of truly critical functions, which alternatives are real.

For those real alternatives, keep enough operational visibility that you could actually move. This does not mean running models on two systems in parallel. It means you understand how Vendor X works, you have people who know it, and you are not blindsided by its roadmap or pricing. If you had to move, you would not be starting from zero.

Be concrete about what that list buys you, because "optionality" is a word that survives a slide review without ever being cashed. Four things get easier, and they are the four things that actually happen to vendor relationships. When your vendor raises prices sharply, you can move in months rather than years, on a timeline that makes the alternative credible in the room where the renewal is negotiated. When they pivot the roadmap away from what you need, and model vendors do, you can evaluate a replacement instead of commissioning a study to find out whether one exists. When a regulator or a customer contract forces a workload elsewhere, you can put it there. And when you divest a business unit, you can sell it without sending the vendor relationship along too. Each is a real event with a date attached, and the list is what turns it from a crisis into a decision.

Hold the line on new systems and leave the old ones alone. New critical infrastructure gets built behind the abstraction and on open standards, trading a little convenience for the freedom to leave. Existing systems already wired deep into the vendor stay put. Do not migrate them for the sake of purity: that is how you pay the full switching cost and buy nothing with it.

Picture this on a real balance sheet. Take an organization with $200M in AI spending. It picks one vendor for model training and invests the time to become genuinely expert. It runs serving through an abstraction it owns. It keeps data pipelines on its existing infrastructure rather than lifting everything onto a model vendor's platform, and builds monitoring on open standards. New systems get the discipline; old ones are left in place. The result is roughly 80–90% of the benefit of full vendor diversity (you can actually move when you need to) without the cost of maintaining deep expertise across five platforms you never fully learn.

No vendor strategy is free. Go all-in on one vendor and you sacrifice some flexibility. Spread across many and you sacrifice efficiency. The honest question is which trade-off fits your organization, and for most enterprises the answer is the same: consolidate on one primary vendor per function, and architect so you are not actually trapped. That buys you optimization now and mobility later. It is not the vendor-diversity answer. It is the pragmatic one.

The Relationship, Not Just the Contract

Architecture manages the cost of leaving. But there is a second half to vendor strategy that architecture does not touch, and most enterprises never work it deliberately: the maturity of the relationship itself.

Most AI vendor relationships are transactional. You buy a platform, you deploy models, you call support when something breaks, and the vendor has no real stake in whether you succeed. A few enterprises have something different: a strategic relationship, where the vendor shares your upside and downside and helps you think through the hard problems. Most enterprises are in the first group and belong closer to the second. They just do not know how to get there. The relationship moves through four levels, and it is worth being able to locate yourself on them.

Level 1 is transactional. You have a vendor, a contract, and an SLA. You pay for the product and they support it to the letter of the agreement. Beyond that you are on your own. The vendor does not know your governance requirements, your risk tolerance, or which of their systems are critical to you. Their incentive is simple: sell more, support to contract. And when something goes wrong, that same incentive works against you: their move is to point at the contract and say *that is not covered.*

Level 2 is advisory. The vendor assigns someone to your account. They understand your use cases, suggest ways to use the platform better, flag new features, help you think through architecture. It is a real improvement (the vendor has context now), but the relationship still runs one direction. You bring problems, they respond. They are not proactively partnering in how you govern.

Level 3 is collaborative. The vendor understands your governance requirements and you talk about them regularly, not only when something breaks. When regulations shift, you discuss the implications together. When new use cases come up, you work through the architecture together. The vendor now has skin in your success: if their product creates a governance problem for you, they help fix it. Few enterprises get here, because it requires the vendor to field people senior enough to understand governance, and it requires you to be open enough to share your governance challenges honestly.

Level 4 is strategic. The vendor is a partner in your governance strategy, not just a supplier to it. When you are building a new capability (automated retraining, fairness monitoring, drift detection), they think through the implications alongside you. They share your risk and your upside: help you deploy faster and more safely and they benefit through your growth and advocacy; cause you problems and they carry part of the cost. Almost no enterprise has this, because it takes years of trust and a vendor genuinely invested in your success.

When I ask an executive team where they sit on this ladder, the answer is nearly always "Level 2, maybe 3." Then I ask them to name the person at the vendor who knows their governance requirements, and nobody has a name until someone offers the account manager, who knows their renewal date and their seat count. That is Level 1 with better catering.

The reasons for that are structural rather than accidental. Transactional relationships are cheaper for the vendor, so their default is to optimize for revenue and support cost, not to understand your governance. Enterprises, for their part, rarely articulate what they need: they just deploy, and if it works, great, and if not, they reconfigure or shop around. Switching costs are high enough that you hold real leverage, but because you never designed your governance around the platform, you do not use it. And most fundamentally, enterprises do not know what "strategic" even looks like, so they cannot ask for it or say what they would offer in return.

Moving up the model is not mysterious, but it is deliberate work. Start with your own non-negotiables. What does the vendor's platform actually need to support for you to govern well: audit trails, explainability, monitoring, integration with your data layer, API access? Write them down as requirements, not nice-to-haves. Then assess the vendor honestly against them. Does the platform support your governance needs and your scaling plans? Is the roadmap headed where you need it? If not, switching may be the right call, but at least now you are deciding deliberately instead of drifting.

If you want to stay, propose the upgrade explicitly. Tell the vendor you want to move from transactional to advisory, and be concrete about what that means: quarterly strategic conversations rather than crisis-only calls; a real exchange of your governance requirements and how their platform supports them; visibility into their roadmap with your input on it. And be equally concrete about what you offer in return. You will stay as you scale, give feedback that makes the product better, and may become a reference customer. As the relationship matures toward collaborative and strategic, that same structure deepens: shared goals, mutual investment, and enough trust that they share their roadmap and you share your constraints.

When you upgrade the relationship, upgrade what you negotiate along with it. The SLA that works for a non-critical system does not work for one touching millions of customers, so negotiate uptime and response terms for critical systems specifically. Negotiate integration rights: whether you can wire their system into your governance infrastructure, access your own data, and build on their APIs. Negotiate custom development for the governance capabilities their platform lacks, and at what price. Negotiate roadmap influence, so that when they are not building something you need, you have a path to reprioritize it. Negotiate for scale, because a portfolio that keeps growing should change the support model and earn you dedicated resources. And negotiate governance collaboration directly: do they field people who actually understand regulation, fairness, monitoring, and incident response?

None of this is free. Strategic relationships are expensive because you are asking the vendor to commit people and time, and they will expect to be paid for it, through higher contract values, revenue sharing, commitments of three to five years, and a claim on their focus. You invest real time too. The payoff is that you deploy faster and more safely, backed by a vendor who understands your business and helps solve problems when they arise. Most enterprises never make that trade because the upfront investment feels steep; the ones that do have run the longer math.

Where you should be scales with exposure, on the same thresholds Chapter 22 uses for governance scale. With 1–5 systems, transactional is fine: you know the platform and can solve your own problems. From 5 to 20, you should be advisory, with someone at the vendor who understands your governance requirements. From 20 to 50, you are too invested to accept anything below collaborative. Past 50, it has to be strategic, because you are carrying portfolio-scale risk on a support contract. Most enterprises sit exactly one level below where their exposure puts them: not because they are doing anything wrong, but because they never negotiated the upgrade. The vendors willing to move up with you are, not coincidentally, the ones worth staying with.

What the Manufacturer Was Missing

Go back to the company that stayed. They did not lack vendor strategy in the boardroom sense: they had slides, a budget line, and a transition analysis with a defensible methodology and a hard number at the bottom. What they lacked was everything that would have had to happen in the five years *before* it for that number to come back smaller.

The integration points had never been inventoried, so the 47 systems were a discovery rather than a fact already on a page. The abstraction had no owner, so applications called the platform directly. On the day each was written, that was the fastest path. There was no shortlist of realistic alternatives, so evaluating a replacement became a research project instead of a decision. Governance requirements existed as folklore rather than as a document, so the vendor had never been measured against them and had no reason to care. And the quarterly review had lapsed years earlier, because the platform mostly worked, and reviews of things that mostly work are the first meeting anyone cancels.

That is the unglamorous half of vendor strategy, and it is the half that decides the outcome. Vendor strategy without the trap is not a smarter contract. It is the patient management of two numbers over time: how much it costs you to leave, and how much the vendor is actually invested in your success.

FIGURE 11.1

The Switching-Cost Iceberg

What you budget to leave a vendor — and what leaving actually costs.

ABOVE THE WATERLINE

The four line items everyone budgets

New licenses	Data migration	Retraining	Parallel running
Contract, seats	Models, configs	New interfaces	Both stacks live

THE WATERLINE WHAT THE ESTIMATE SAW

BELOW THE WATERLINE

The five layers that surface only once you commit

1	**Architectural dependencies**	47 downstream systems assumed the old API.
2	**Data-format translation**	Every consumer of the output format, re-tested.
3	**Performance characteristics**	Queues and caches tuned for the old system.
4	**Failure modes**	Vendor B fails differently. You learn in prod.
5	**Organizational knowledge**	The quirks and workarounds leave with the team.

≈ 3× the estimate WHAT LEAVING ACTUALLY COSTS

One manufacturer priced its exit at $8M and 18 months.
They stayed. The lock-in was never in the contract.

Workshop use: teams fill the four budgeted items, then price each hidden layer.
If the honest total is not roughly 3× the estimate, the estimate is not finished.

FIGURE 11.1

The Bottom Line

Vendor strategy is not a contract decision or a headcount of how many vendors you use. It is the deliberate management of two things the contract barely touches: the true cost of leaving, and the maturity of the relationship. Switching costs run roughly three times the visible estimate because most of them are hidden architectural debt (integration points, data formats, failure modes, and lost institutional knowledge that surface only after you commit), so you price the exit during selection, not during the transition. Diversifying your vendor stack does not buy flexibility; it buys fragmentation, because lock-in is a problem of abstraction, not of vendor count, and the fix is architectural: one strong primary vendor per function, wrapped in an abstraction layer you own. On top of that architecture, run the relationship up the maturity ladder from transactional to strategic in step with how many systems you have staked on the platform. Do both and you get optimization now and mobility later — and the enterprises that win this rarely switch vendors at all; they simply keep the power to, and a partner invested enough to make staying worth it.

Field Kit

1. The switching-cost estimator (five hidden layers to price before you sign). For any vendor you are selecting or considering leaving, size each layer explicitly — none of them appears in the standard migration budget: - *Architectural dependencies* — how many downstream systems assume this vendor's API behavior, error handling, and data flow? - *Data format translation* — how many consumers are built around its representation of outputs, uncertainty, and feature importance? - *Performance characteristics* — how much rate-limiting, queuing, and caching is tuned to its latency and burst profile? - *Failure modes* — how much of your error handling is shaped by how *this* vendor fails? - *Organizational knowledge* — which undocumented workarounds leave with the people who built them?

Then do the arithmetic in the right order. Pad your slowest visible piece by 30% — usually re-abstraction and integration, rarely data migration — then add the five layers *on top* of that padded figure. The pad is not the answer; the layers are, and together they take a careful estimate to roughly three times your first number.

2. The one question to bring to every vendor-selection meeting. *If we needed to move off this vendor in three years, how hard would it be — in dollars and in* *months?* If the answer is "very hard," negotiate better terms up front or choose differently. Ask it before you sign, never after.

3. The fragmentation check. Before adding a second vendor "for flexibility," confirm you are buying flexibility and not fragmentation: Can the two vendors' functions actually be swapped or recombined, or are you just running two systems that don't talk to each other? If moving would be exactly as hard afterward, you have added complexity, not optionality — solve it with an abstraction layer instead.

4. The five-move vendor architecture. (1) One primary vendor per critical function, chosen best-in-class. (2) An abstraction layer you own between your systems and that vendor, built as Chapter 10 describes. (3) A named, specific list of realistic alternative vendors per critical function — names, not categories. (4) Enough operational visibility on those alternatives that you could move: people who know them, no surprises on roadmap or price. (5) New critical systems built behind the abstraction; existing deeply integrated systems left alone.

5. The relationship-maturity diagnostic. Locate each critical vendor on the ladder — Transactional, Advisory, Collaborative, Strategic — then check it against your system count, on Chapter 22's bands: 1–5 systems, transactional is fine; 5–20, advisory; 20–50, collaborative; 50+, strategic. Test your answer by naming the person at the vendor who knows your governance requirements. If you cannot name one, you are a level lower than you think — and you have leverage you are not using. Open the upgrade conversation.

6. The upgrade term sheet. When you propose moving up a level, negotiate the six things that actually change with maturity: uptime and response SLAs scaled to system criticality, integration rights into your governance infrastructure, custom development for governance gaps, roadmap influence, a support model that scales as your portfolio grows, and explicit governance collaboration (do they field people who understand regulation, fairness, monitoring, and incident response?). Bring what you will trade in return — scale commitment, feedback, advocacy — written down before the meeting.

PART V

Operations: Monitoring, Failure & Resilience

CHAPTER 12

Monitoring & Observability for AI

Part V · Operations: Monitoring, Failure & Resilience

"So we'll monitor the model, right?" Everyone nods. The meeting ends. And the next meeting is spent discovering that no two people in the room meant the same thing by it.

When your data engineer says they're monitoring a system, they mean something precise: is it up, is it responding, are queries returning inside their SLA, is storage growing in a way it shouldn't. When your data scientist says they're monitoring a model, they usually mean: is performance degrading, has the input distribution shifted, are we still inside the thresholds we set at deployment. Both are legitimate. Neither is the question that matters most, and the gap between them is where models fail without anyone noticing.

Here is the thing nobody warns you about when you stand up your first AI system in production. A model does not fail the way a database fails. A database that breaks tells you it broke: the query times out, the disk fills, the alert fires, the pager goes off. An AI system that breaks can keep every light green and go on making worse decisions every single day, and the monitoring you inherited from your infrastructure team will not see it. Uptime perfect. Latency clean. Resources healthy. And the model, six months in, systematically wrong in a way that shows up nowhere on the dashboard.

This is the gap that most enterprise AI governance never notices it has. Your security and operations teams know how to tell you a system is healthy. They do not, for the most part, know how to tell you a model is making good decisions. Those are different problems with different solutions, and the confusion between them is where the real risk lives.

Here is the reframe this chapter is built on.

The myth: if uptime is clean and drift metrics are stable, the model is doing what you deployed it to do.

The reality: system health and behavioral health are different problems, and a model can keep every light green while its decisions walk steadily away from what you approved at launch.

The System That Was Perfectly Healthy and Quietly Wrong

Let me give you the case that made this concrete for me.

A financial services company deployed a system to recommend investment allocations. The model was good — genuinely good. It predicted portfolio performance accurately on historical data, and in production it did everything a well-run MLOps shop is supposed to do. It monitored drift: was the model's performance degrading? It monitored feature distributions: was the incoming data different from what the model trained on?

For three months, everything looked fine. No drift. Data distributions stable. The model performing exactly as designed. By every metric on the dashboard, this was a success story.

Then in month four, a compliance officer pulled a sample of the system's actual recommendations to look at them by hand. The allocations were technically sound — nothing that would fail a rule. But they were increasingly concentrated in a narrow set of holdings. The model wasn't breaking. It was drifting *behaviorally*, making steadily more concentrated recommendations without anyone realizing it, because concentration wasn't a thing the monitoring was built to watch.

They caught it because a human looked at what the system was actually deciding. Not because anything flagged. If that compliance officer had been busy that month, the concentration would have kept compounding.

That is the pattern, and once you see it you see it everywhere. The guardrails that fail first in production are almost never the ones where the model broke. They're the ones where the model kept running beautifully and nobody was watching the right signal. I know of a system that stopped working while the model itself was fine: an upstream data pipeline was silently modified and suddenly every feature came through null. The model kept running. It kept producing predictions (confident, well-formatted, completely meaningless) for weeks, because no one was monitoring whether the inputs were even present. I have seen models that minimized their loss function beautifully and maximized customer churn, because they'd learned a correlation that was mathematically true and operationally disastrous. In each case the technical monitoring was green. In each case the system was, by the only definition that matters, failing.

System Health Is Not Behavioral Health

The clean way to hold this distinction is to separate two kinds of monitoring that most organizations treat as one.

System monitoring asks: Is the system running? Is it available? Is it processing requests? Are resources adequate? This is what your operations and security teams already do, and they are usually excellent at it. If your model API goes down, you will know in seconds.

Behavioral monitoring asks a different set of questions entirely: Are the model's decisions changing? Do they still make sense? Are some groups getting systematically different treatment than others? Is the model still producing what you designed it to produce? None of these show up in uptime or latency. A system can be flawless from a system perspective and steadily degrading from a behavioral one.

The reason this matters so much is that the second kind of failure is invisible to the first kind of monitoring by construction. A model can hold its accuracy exactly where it was on launch day (no drift the dashboard can see) while the *distribution of its decisions* walks steadily away from where it started. The allocation model above did precisely that: accuracy intact the whole way, concentration compounding underneath it. Nothing about that failure lives in a number your MLOps tooling reports, because the tooling was built to ask whether the model is still right, and the question that mattered was what the model is now doing. Statistical accuracy and good decisions are not the same measurement, and a model can hold one while losing the other.

I watched this play out with a company that deployed a model to prioritize customer support tickets. The system monitoring was pristine: the model generated a prediction for 100% of incoming tickets, the infrastructure was solid, nothing ever paged. But six months in, the model had started systematically de-prioritizing tickets from customers in certain regions. Its accuracy was stable: it was still, technically, predicting urgency correctly by its own definition. The *decision pattern* had shifted: genuinely urgent tickets from one region were being treated as less urgent. Nobody noticed, because outcome review happened only in infrequent business meetings and behavioral monitoring didn't exist. They finally caught it through a customer complaint — which is to say, they caught it after the customer did. By then they had months of bad decisions to remediate.

The system was set up to catch a model that *fails to produce predictions*. It was not set up to catch a model that *produces degraded predictions*. Almost every enterprise I see has built the first and skipped the second.

The Signals That Actually Matter

If technical drift and feature distributions aren't the signals that catch the real failures, what are? In practice, five things need watching in production, and only two of them tend to be instrumented.

Signal integrity. Is the model receiving the data it expects (not just in the right schema, but from the same sources), and are those sources themselves stable? This is where the null-feature failure hides. The schema was valid. The pipeline had changed underneath it. Nobody was checking whether the input signal was actually present, only whether it was well-formed.

Decision consistency. When the model makes a decision, is it consistent with what you'd expect? This one is stubbornly qualitative: you cannot fully automate it, and that is exactly why organizations skip it. But someone has to periodically look at real decisions the system is making and ask, plainly, *does this look right?* I saw a recommendation system that was technically working fine (every metric green, no drift, no alerts) while it had learned to serve steadily more extreme recommendations to a small segment of users. No instrument caught it. Someone read a page of actual recommendations and caught it immediately. That is how the concentration problem surfaced too, and half the failures I've described: a human looking at outputs found in an afternoon what the metrics missed for months.

Outcome correlation. You built the system to predict X and optimize for Y. In production, are the outcomes you're getting actually consistent with that intent, or has the model found a pattern that moves the metric up while moving the outcome the wrong way? The churn-maximizing loss-minimizer is the cautionary tale here. The math was flawless. The business result was the opposite of what anyone wanted.

Distribution shift that matters. Most systems monitor whether *any* input distribution has changed. But not all shifts matter equally. A shift in the age distribution of your customers might be critical; a shift in their geographic distribution might be irrelevant — or the reverse. Monitoring "something changed" generates noise and false confidence in equal measure. You want to monitor shifts *relative to what you're trying to predict,* not shifts in the abstract.

The rare cases. Your model works great on the common case and often terribly on the rare one. And the rare one is frequently the one that matters most. Aggregates hide segments, as established; here the segment is the 2% of cases that carries most of your risk, and it is small enough that the system can fail on all of it without moving the headline number a point. Track the rare case as its own line, with its own threshold, or you will not see it break.

Notice what these five have in common. The organizations that are ahead on this are not the ones with the most sophisticated monitoring platforms. They're the ones who are clear about which signals actually matter for their use case, and those signals are usually nowhere near the standard model-monitoring dashboard.

Four Layers, and Where Everyone Stops

The cleanest way I've found to structure this is to build monitoring in layers, and to be honest with yourself about which layer you actually have.

Layer 1: Infrastructure. Is the system up, and is it fast? Table stakes. Use the same application and database monitoring you already run for everything else.

Layer 2: Technical drift. Is model performance degrading? Is the input distribution moving? Important, and well served by standard MLOps tooling. Most mature teams have this.

Layer 3: Decision monitoring. What is the system actually *deciding,* and is the decision distribution consistent with what you expect? This almost always requires custom instrumentation, because no off-the-shelf tool knows what "a reasonable decision" means for your business.

Layer 4: Outcome monitoring. Are the outcomes you're getting in the world aligned with what you set out to achieve? This means tracking real business outcomes, not the system's own metrics; it is usually the hardest layer to build and the most important one to have.

Here is the uncomfortable summary of nearly every enterprise AI program I encounter: Layer 1 and Layer 2 are solid. Layer 3 and Layer 4 are weak or entirely absent. And that is exactly where the guardrails fail, because Layers 1 and 2 are about *the system*, while Layers 3 and 4 are about *what the system does in the world*, and those are different problems that no amount of infrastructure monitoring will solve for you. So when someone tells you "we're monitoring the model," the useful follow-up is: which layer? If the honest answer is "technical drift," you are watching the two layers where the dangerous failures don't live.

Guardrails Can Only Protect What You Can See

There's a fashionable answer to all of this, and it's worth taking apart because it's half right. Build guardrails, the argument goes — a layer that watches your AI's output and blocks the bad outcomes: hallucinations, bias, toxic content, out-of-distribution predictions. The concept is sound. The implementation usually fails, and it fails for a reason that has nothing to do with the guardrail logic.

When a guardrail deployment disappoints, the story is almost always the same. The team deploys the system, wraps the output in guardrail rules, flags what violates them, and declares the system safe. The guardrail catches some problems. It misses others because it cannot see upstream. Problems appear it was never designed to catch. And the enterprise concludes that guardrails don't work — when what actually failed was the observability *underneath* the guardrail.

To see why, walk the layers a request actually passes through. There's the **input**, the raw data fed in. The **processing**: feature engineering, preprocessing, normalization. The **model** itself. The **raw output**, before anything is cleaned up. The **post-processing**: thresholding, formatting, business logic. And finally the **guardrail**, at the very end of the line. Guardrails operate at that last stop. They can inspect the final output. But if the problem originated upstream (garbage input, a drifted preprocessing constant, the model wandering into an edge case), the guardrail sees, at best, a symptom, and never the cause.

This is where the blind spots live, and they compound. **Input blindness:** most enterprises can tell you a request arrived, but not whether the input was in the expected range or looked anything like the training data. **Processing blindness:** preprocessing is treated as deterministic and therefore trustworthy (you ran the code, it did its job), so when a normalization constant drifts or a feature-engineering step hits an edge case, nobody notices until the damage is large enough to surface downstream. **Model blindness:** accuracy is checked retrospectively, once ground truth arrives; real-time visibility into whether the model is confident, uncertain, or far outside its training distribution is rare. **Output blindness:** the raw output distribution goes unwatched entirely, because everyone jumps straight to the final, post-processed result.

Stack those blind spots up and the guardrail's limits become obvious. When the input shifts, the guardrail can flag "model not confident," but it cannot tell you *why*: you don't know the input moved. When preprocessing drifts, the guardrail catches abnormal output, but you'll spend a day debugging the model when the fault was two layers upstream. When the model hits an edge case, the guardrail says "output flagged" and leaves you to guess which edge case did it. A perfect guardrail on top of a blind system is a smoke alarm with no way to tell you which room is burning.

So the sequence enterprises want is backwards. They build the protection layer before the visibility layer (guardrails first, observability maybe never), and then something goes wrong and they don't have the telemetry to understand it. They know a guardrail triggered. They don't know whether it was an input problem, a model problem, or a preprocessing problem, and they get to retrace the last hour of logs to find out. Observability is not the thing you add after the guardrails prove insufficient. It is the foundation that makes the guardrails work at all.

Nobody Owns Behavioral Health

Everything I've described so far can sound like a tooling problem, and it is comforting to treat it as one, because tools you can buy. But you can build every dashboard in this chapter and still fail, because the hard part was never the technology.

The hard part is organizational, and it comes down to a single unanswered question: *who owns behavioral health?* Not system health: that's the operations team, and they've got it. Not model validation at deployment: that's the data science team, and they did it once. Who is responsible, continuously, for knowing whether the model is still making good decisions this week?

Watch where the failure lands and you'll see it's always at an organizational boundary. Operations owns system monitoring and is excellent at it. Data science owns validation, which happened once and was never refreshed. The business owns outcomes, reviewed infrequently and at high altitude. Behavioral fairness monitoring is missing or ad hoc. Every individual team is doing its job. The gap between the jobs is where the model degrades for six months undetected. If the answer to "who owns this" is *the data science team*, then they need real time and budget for continuous monitoring. They cannot build the model and hand it off. If the answer is *the operations team*, they need to understand that model monitoring is a genuinely different discipline from system monitoring and be resourced to learn it. If the answer is *nobody*, which is the most common answer, then you have already bought yourself the invisibly degrading model. You just haven't met it yet.

Most organizations avoid making this decision at all, which is precisely why most organizations have the gap. Observability is plumbing that sits alongside the system and watches, and building it upfront costs you a few weeks nobody will congratulate you for. So teams build the model, deploy the guardrails, and assume they're finished. They've built the protection and skipped the visibility, and the bill for that arrives later, with interest, in the form of a failure they cannot diagnose.

What Actually Works

Monitoring tells you a decision changed. Whether that change mattered is a measurement question, and Chapter 18 is where the metric gets chosen and given an owner.

Strip away the tooling debate and the maturity-model diagrams, and continuous behavioral monitoring comes down to a handful of practices any team can start on immediately.

Establish baselines before you deploy. Understand what "normal" looks like while you still can: the accuracy distribution, the outcome rates, the decision distribution across your important segments. That baseline is the reference point that makes every later comparison meaningful. Without it, you can measure change but you can't recognize it.

Measure the right things continuously. Track, as continuously as you track system health, the metrics that actually matter: accuracy and its distribution *across segments* (not just in aggregate, where the segment failures hide), outcome rates (how often the recommended action actually succeeds), decision distribution (is the treatment or recommendation rate shifting?), and fairness metrics (is the outcome distribution stable across the groups you care about?). Continuous is the operative word. A quarterly check will find the problem a quarter late.

Alert on meaningful change, not on noise. Accuracy will wobble day to day, and that's fine. Accuracy *systematically degrading* is not fine, and neither is a decision rate walking away from its baseline. The whole discipline lives in defining "meaningful" for your use case: a 1% drop in one segment might be noise, a 5% drop in another might be a crisis. Get that definition wrong toward sensitivity and you train your team to ignore the alerts; get it wrong toward tolerance and you miss the drift you built the whole system to catch.

Treat behavioral failures like incidents. When something goes wrong, run the same postmortem you'd run for an outage. What changed in the environment? In the data? Was the model's behavior actually appropriate given that change? Would monitoring have caught it earlier, and if not, what instrument do we add so it catches the next one? A model degradation is an incident. Give it the same seriousness, and the same learning loop, that you give a system one.

And if the whole thing feels too big to start, don't boil the ocean. Pick your single highest-impact model. Choose three or four behavioral metrics for it: your primary success metric, something about equity, something about the actual business outcome the model exists to move. Put them on a dashboard. Look at it every week. When something moves, understand why before you move on. That's it. That is behavioral monitoring. It is not complicated, and that is exactly the point: it's not hard to understand, it's hard to *own*, week after week, when nothing appears to be wrong.

The enterprises that handle AI safely are not the ones with the best models. They're the ones with the best visibility into what those models are actually doing: where the data is coming from, what's happening at every layer from input to final output, and whether the decisions this week still look like the decisions you signed off on at launch. The systems that fail without ever tripping an alert are never the ones where the model broke. They're the ones where nobody was watching the right signal.

FIGURE 12.1

System Health vs. Behavioral Health

Five monitoring roles, filled twice — once for a database, once for a model.

THE SAME ROLE	**System health** What a database monitor sees	**Behavioral health** What an AI system also needs
Headline metric	**Uptime** 99.98% this quarter. Every health check answered.	**Segment drift** 22% of volume is now a population it never saw.
Early warning	**Latency** p99 held under 180 ms all month. Nothing alerted.	**Decision-mix shift** Approvals moved 61% to 74%. No rule changed. No retrain.
Correctness test	**Error rate** 0.02% of calls error out. Returning is being right.	**Confidence calibration** "90% confident" is right 71% of the time. Nothing errored.
Strain signal	**Throughput** 40,000 requests a minute. Queue depth flat all month.	**Override rate** Reviewers reverse 1 in 3. Last quarter it was 1 in 20.
Who provides it	Ships with the platform.	You build it, or nobody does.

All four metrics on the left stayed green the entire time.

A database monitor is not broken when it reports this. It is answering a different question — whether the system ran, not what it decided.

Workshop use: fill the right column for one live model, row by row, with real numbers. Any row you cannot fill is a row nobody is watching — assign an owner before you leave.

FIGURE 12.1

The Bottom Line

Monitoring an AI system is not monitoring a database. A database that fails tells you; a model can keep every light green while it makes steadily worse decisions for months, and the system monitoring you inherited will never see it. The distinction that matters is between *system health* (is it up, is it fast) and *behavioral health* (are the decisions still good, are some groups being treated differently, are the outcomes still what you intended), and almost every enterprise has built the first and skipped the second. Guardrails don't fix this, because guardrails sit at the end of the pipeline and can only catch what upstream observability lets them see. The work that actually protects you is not technically hard: baseline the behavior, watch the decisions and outcomes continuously, alert on meaningful change, and, above all, put one name against behavioral health specifically, because operations already owns system health, data science already owned validation, the business already owns outcomes at quarterly altitude, and the model degrades in the gap between them.

Field Kit

1. The "which layer are you monitoring?" question set for your next review. Bring these five questions to any meeting where someone says "we're monitoring the model": - Are we monitoring the *system* (uptime, latency) or the *model's decisions*? Show me the second dashboard. - Which of the four layers do we actually have — Infrastructure, Technical drift, Decision, Outcome — and which are we missing? - When did a human last look at a sample of the system's actual decisions and ask "does this look right?" - If a group were getting systematically worse treatment, which instrument would catch it, and how long would that take? - Who, by name, owns behavioral health for this model, continuously — not who validated it at launch?

2. The four-layer coverage check. Walk the list and mark a layer *covered* only if you can name the dashboard and the person who reads it. Anything you have to argue for is missing. - Layer 1 Infrastructure: uptime, latency, and resource monitoring are in place. - Layer 2 Technical drift: performance and input-distribution drift are tracked with MLOps tooling. - Layer 3 Decision monitoring: the decision distribution is instrumented and reviewed against expectation. - Layer 4 Outcome monitoring: real business outcomes (not model metrics) are tracked back to the model's decisions.

Then say which layer is missing, out loud, in the room. If it's 3 or 4 — and it nearly always is — the layers you do have will go on reporting that everything is fine right through the failure you care about, and your remediation window is however long it takes a customer to complain.

3. The five signals worth watching (audit each of your top models against this list). Signal integrity (are inputs actually present and from stable sources?), decision consistency (does a human periodically sanity-check real decisions?), outcome correlation (are we getting the outcome, or just moving the metric?), distribution shift *that matters* (relative to what we predict, not shift in the abstract), and rare-case performance (are we tracking the 2% that carries the risk, or averaging it away?).

4. The pre-guardrail observability checklist. Before you trust a guardrail, confirm you can see each layer upstream of it: input (distribution, anomalies, drift since deploy, sample inputs), processing (feature correctness, ranges, drift, edge cases), model (prediction distribution, confidence, uncertainty on out-of-distribution inputs), output (raw vs. post-processed, distribution), and the guardrail itself (which rules fire, false-positive rate, coverage). Any layer you can't see is a failure the guardrail will flag without being able to explain.

5. The behavioral-monitoring starter, for one model this month. Pick your highest-impact model. Choose three or four metrics: the primary success metric, one equity metric (decision or outcome rate across groups), and one business-outcome metric. Baseline them before or at deployment. Put them on a dashboard reviewed weekly, with alerts tuned to *systematic* change rather than daily noise. When a metric moves, run a postmortem and add the instrument that would have caught it sooner.

6. The ownership decision, forced. For each production model, write down the name that owns *behavioral* health specifically — not system health, which operations already has, not launch validation, which data science already did, not quarterly outcome review, which the business already runs. If that name has no continuous-monitoring time in a budget, you have an intention rather than an owner; if there is no name at all, treat it as an open incident, because the invisibly degrading model is already on its way.

CHAPTER 13

Designing for Failure

Part V · Operations: Monitoring, Failure & Resilience

You are sitting across from an operations leader whose AI system is working perfectly. It performs exactly as designed. It is also wrecking their business.

The system makes a recommendation every hour for a critical workflow decision, and it is right about 92% of the time—significantly better than the humans it replaced. But the workflow around it was built on an assumption nobody ever wrote down: that a human would always be there to catch the 8% when the machine got it wrong. Now the 8% is happening with no human in the loop. There is no fallback. There is no way to pause and escalate. The system keeps recommending, the business keeps following, and the outcomes keep getting worse until someone finally notices the trend and asks what changed.

Nothing changed. The model is doing precisely what it was validated to do. That is the entire problem.

This is not a model failure. It is a design failure, and it is the most expensive kind of failure in enterprise AI because it does not look like a failure at all. There is no crash, no error, no pager going off. The metrics you are watching are green. Somewhere underneath them, the business is bleeding.

Here is the reframe this chapter is built on.

The myth: an AI system is safe once the model is accurate enough, because an inaccurate model is the failure you need to guard against.

The reality: accuracy is about the model being right; resilience is about the system continuing to work when the model is wrong, and most production failures are design failures, not model failures.

Most organizations pour their attention into accuracy and almost none into resilience. They treat an AI system as if it has exactly one failure mode (the model is inaccurate), and they build their entire operational posture around guarding against that one mode. Then they get blindsided by the dozen other ways a production AI system fails, most of which never trip a traditional alarm.

The Model Is Rarely the Thing That Breaks

Start by noticing how many of these disasters have nothing to do with the model being bad.

One organization built an AI system to route support tickets to the right team. It worked beautifully until work volume spiked. Suddenly tickets were being misrouted to teams that were already underwater, creating cascading delays. The model did not fail. The system failed, because there was no surge protection: no way to throttle input when the pipeline could not keep up, no circuit breaker. It just kept processing, making a bad situation worse.

Another integrated AI into their hiring pipeline to screen résumés. The model was well-built, validated, and tested. Then, halfway through a hiring season, they realized they had never asked what happens when the model is confidently wrong about candidates from a particular geographic region. It had been rejecting qualified people systematically for weeks. When the model failed, there was no recovery mechanism, because recovery had never been part of the design.

A third built a system to predict customer churn and automatically trigger retention actions. It worked for six months. Then the retention actions themselves began changing customer behavior in ways the model had never seen. The model started predicting from patterns that no longer applied, and the system cascaded, firing off increasingly aggressive retention offers because churn appeared to be getting worse, which made it worse.

These are all preventable disasters, and none of them are model problems. The model could be right while the training distribution has shifted underneath it. The model could be confidently overconfident. The system could be amplifying its own errors through a feedback loop. The upstream data could have changed. The business context the model was trained for could have evolved out from under it. Every one of these requires thinking about the whole flow (inputs, processing, decision, action, feedback) rather than obsessing over the single node in the middle where the math happens.

The organizations that do this well do not separate the AI from the business process. They design the AI as one component of a larger resilient system. That system has fallbacks. It has observability. It has circuit breakers. It has escalation rules. And it gets tested not just for whether it is accurate, but for how it behaves when it is not.

Three Kinds of Failure, and the Two You Are Not Watching

Most enterprises have spent years building operational discipline around traditional software failures: crashes, latency spikes, data corruption, security breaches. These failures are discrete, observable, and loud. When one happens, everyone knows immediately. The pager goes off, you hold a post-mortem, you fix it. AI systems fail differently, and it helps to sort those failures into three types, because most organizations are built to catch only the first.

Type One is the failure you already monitor for. The service goes down. The API returns 500s. Inference latency spikes to ten seconds. The model crashes under load. A recommendation model's serving infrastructure runs out of memory, the API starts erroring, monitoring alerts, you page the team, and you either restart the service or roll back the version. This is standard operational hygiene, and you almost certainly have it covered. It is the easy case, and it is the one everyone over-invests in because it is familiar.

Type Two is the failure that keeps your compliance people awake. The system is up. It is returning predictions. It is throwing no errors. But the predictions are wrong or systematically misleading—and you will not know for hours, days, or longer. Go back to the churn system from a few pages ago. Its predictions never stopped arriving and not one of them came back malformed. What happened was that the retention offers it fired taught customers to wait for a discount, the model read the resulting behavior as rising churn risk, and the trigger rate climbed from eleven percent of the base to twenty-six over five months. Latency was normal. Throughput was normal. Every dashboard the team watched stayed green while the system discounted its way through a quarter of the customer base at collapsing margin. Nobody in engineering caught it. A finance analyst reconciling discount spend caught it, four months in. Bias is one flavor of this (the zip-code credit model from Chapter 4 is the same shape), but Type Two is much broader than fairness. It is any wrong answer that arrives well-formed, on time, with a 200 status code.

Type Three is the rare, terrifying one: the model that passed testing and fails in production. A fraud detection model trained on six months of historical transactions tests at 94% accuracy. You deploy it, and the real world turns out to be different enough that accuracy drops to 68%, and you do not notice for weeks. Maybe a feature engineered correctly in the test pipeline is computed differently in production. Maybe the feature store returns different values in different environments. Maybe fraud simply evolved, and the model has never seen the new pattern because training was on a world that no longer exists. Type Three failures are the hardest to defend against, because defending against them requires admitting that your test environment and your production environment are fundamentally different systems. Most organizations assume they are identical until proven otherwise.

Here is the uncomfortable arithmetic. If you are guarding only against Type One, you are exposed to Type Two and Type Three, and those two are both more common and more damaging than the dramatic crash you have rehearsed for. Most operational incidents in AI are not the pager-off-at-midnight kind. They are the silent kind, where the system is working perfectly by every metric you monitor while producing wrong answers for real people.

The good news is that Type Two is the one you can start on immediately, and it will catch the most common failures. It does not require exotic technology. It requires three things you can build with the tools you already have:

- **Baseline distributions.** Decide what your model's outputs are *supposed* to look like. For the churn system: the expected retention-offer trigger rate, and its expected spread across tenure bands, plan tiers, and account sizes.
- **Continuous measurement.** Measure, on a tight cadence, whether the actual distribution still matches the baseline. If the trigger rate drifts from 11% to 14%, that is an alert. By 26% you are no longer investigating a drift, you are cleaning up a quarter. If one plan tier moves while every other tier holds steady, that is an alert.
- **Cohort monitoring.** Slice by sensitive attributes and track whether accuracy and fairness stay stable across cohorts. For anything that touches a protected class, this is not optional.

Type Three demands something harder than technology: honesty. Validate the model against a hold-out set drawn from *production* data—not a month later, a week later.

Compare feature distributions in production against training; a feature that was normally distributed in training and shows up bimodal in production means you are living in a different world than you thought. And build a rollback plan, because if production accuracy is materially worse than test accuracy, you need to be able to retreat fast. Most organizations never build the rollback, because they have assumed they will never need it.

What Happens at 2 A.M.

Detection tells you the house is on fire. It does not put the fire out. For that you need to have decided, in advance and in daylight, what the system does when it cannot do its job.

A financial services company's recommendation engine went down at 2 A.M., not catastrophically, just an inference timeout that cascaded through the stack. Their monitoring was good; they caught it within ten minutes. And then they discovered they had no fallback strategy. What should the application show when the recommendation engine is gone? Generic recommendations? Yesterday's recommendations? Nothing at all? The team had never had the conversation. They had built a system that could fail and had no plan for the moment it did. It took them forty-five minutes to decide, implement, and redeploy a fallback. For thirty-five of those minutes, their application was either broken or silently degraded, and no human had ever made a deliberate choice about which.

This is the typical enterprise pattern, and it is worth naming plainly: build the system, deploy it, hope it does not fail, and when it fails, decide what to do in real time. That pattern resolves into one of four ugly outcomes: a cascading failure that takes the whole application down; a silent failure that returns a null or default the application cannot distinguish from a real prediction; a slow failure that hangs waiting for a response; or an inconsistent failure where different parts of the system break in different ways and each downstream consumer improvises its own handling. None of these are acceptable for anything that touches customers, operations, or compliance.

Notice, too, what actually fails first. It is rarely the model. Models are generally robust once they are in production. What fails is the infrastructure around the model: the serving platform, the data pipeline that stops delivering fresh inputs, the API that develops latency spikes, the inference process that consumes more memory than budgeted, the downstream system that stops accepting output because a format or an endpoint changed. The effect is identical regardless of cause: your AI system is no longer functional, and the application calling it still needs an answer.

A system that can fail safely needs three components, and only one of them is technical.

Detection means knowing when the system is down or degraded, not "the model didn't return a prediction," which might be perfectly normal if it wasn't confident enough, but "the system is not functioning within its SLA." That requires health checks that ask the real question: is the service returning predictions within the time window, at the confidence level, that we defined as healthy?

A fallback strategy is a business decision wearing a technical costume. When the AI is unavailable, does the application serve yesterday's recommendations, a generic default, a degraded interface, or a handoff to a human? The right answer depends entirely on context. A recommendation engine can fall back to "most popular." A fraud system falls back to "hold the transaction for manual review." A pricing system falls back to yesterday's prices or a safe default. The point is not which answer you choose. The point is that a human chose it deliberately, in advance, rather than inventing it at 2 A.M.

Activation and rollback is the part everyone skips. A fallback is not a plan document; it is a runbook with explicit steps. Who gets notified? What command flips the system into fallback? How do you know when the AI is healthy again? What does the return to normal look like? How long are you willing to run degraded before that itself becomes an incident? And (this is the part that separates the organizations that survive from the ones that panic) the runbook has been practiced. The team has run through it in a test environment and watched what breaks, because you always discover something you misunderstood about your own fallback the first time you actually pull the lever.

If you want to know why this work gets skipped, the reasons are honest and a little embarrassing. First, it is depressing: when you are excited to ship, planning the funeral feels like pessimism. Second, it is operational grunt work (write the runbook, test it, update it) with no architecture puzzle in it to hold anyone's attention. Third, there is a seductive misconception that good monitoring makes fallback unnecessary. It does not. Good monitoring means you catch the problem in two minutes instead of twenty. But with no fallback, those two minutes are still two minutes of broken service, and the clock does not stop just because you saw the fire early.

There is one more subtlety worth planning for. The most common scenario is not "the system is completely down." It is "the system is slow" or "the system is returning low-confidence predictions." For those, you often do not need a full failover: you need graceful degradation. Return fewer recommendations. Widen the scoring threshold. Keep the system running at reduced capability rather than dropping it entirely. Designing for degraded operation, not just total failure, is what keeps a bad day from becoming a bad week.

The Framework That Prevents Disasters Versus the One That Documents Them

By now you may be nodding along and thinking your organization already has all of this. You have an operational risk framework. You have policies, documentation, and processes. Here is the test that reveals whether any of it is real: ask your team to walk you through what happens when a model breaks. Not what is *supposed* to happen—what *actually* happens.

The answer is almost always some version of: "We have a process documented. It has never been tested. Nobody really knows if it works. When there's an actual incident, we figure it out as we go." That is not an operational risk framework. That is documentation, and documentation is static in exactly the moment you need something dynamic.

The difference is the difference between two sentences. Documentation says: "Here's what we do if a model fails." A framework that prevents disasters says: "Here's how we detect that a model is failing. Here's how we communicate. Here's what we do in the first minute, the first hour, and the first day. Here's what we measure to know whether we're recovering. Here's how we learn from it when it's over." The first is a filed artifact. The second is a discipline that has been rehearsed under stress. Frameworks that actually prevent disasters share five traits.

Detection before impact is the foundation. You have to know something is wrong before your customers or your business does. Monitoring for API errors, latency, and crashes is necessary and insufficient; for AI you also monitor accuracy degradation, input distribution shift, fairness drift, output distribution change, edge-case failures, and cost spikes. Building this is tedious. It is also the load-bearing wall: you cannot prevent a disaster you do not learn about until it is too late.

Escalation that is tested and fast. When the alert fires, what happens next? In most organizations: the alert fires, someone eventually notices, they message the team, they start investigating, and hours pass. In organizations that prevent disasters: the alert fires within thirty seconds, the on-call engineer is paged, within two minutes they have enough context to call it either "turn it off now" or "investigate further," and within five minutes the escalation decision is made. Write those clocks down, and write down with them the three things that make them achievable: alerting clean enough not to cry wolf, an on-call rotation that genuinely functions, and a named person with standing authority to disable the model at 2 A.M. without convening anyone. Name that person in the runbook, by role and by pager, before you need them.

Graceful degradation for critical systems. Some models cannot simply be switched off, because switching them off breaks the whole system. Those need a fallback designed and tested before the crisis: a simpler and more robust model, rule-based logic that covers the common cases, or a default behavior that is safe even if suboptimal. You want to enter every disaster knowing you have an option that is not "the system is broken."

Communication protocols that actually work. When something breaks, people need to be told—quickly, not eventually. For critical models that means internal notification within fifteen minutes (engineering knows, leadership knows) and external notification within an hour if customers are affected. Most organizations do not have these timelines; they have "we'll communicate once we understand the problem," which stretches into hours or days while trust erodes.

A post-incident process that actually learns. Not a post-mortem that gets written and filed, but a loop that understands what happened, names the root cause, identifies what would have prevented it, assigns an owner to build the fix, and tracks it until it is done. The organizations that skip this, or that write the post-mortem and never ship the fix, are the ones where the identical failure recurs six months later, wearing a slightly different hat.

You do not build any of this top-down. You build it through incident response. Your first real incident (a model breaks and causes genuine impact) teaches you what your actual process is, which is usually chaotic and full of gaps. Then you deliberately design the process you wish you'd had, document it, test it, and train the team. And then, crucially, you treat it as something to be *practiced*, not merely filed.

That practice has a name: the incident simulation. Every quarter, you run one. "Pretend this model's accuracy dropped 20% overnight. What happens?" The team walks the process, discovers where it breaks, and fixes it. Next quarter, a different failure mode: a latency spike, a cost explosion, an output distribution shift, a rising false-positive rate. The team gets faster because it has rehearsed. If your team cannot walk through a simulated incident in about thirty minutes, your escalation process is not ready. If they cannot tell you, within fifteen minutes, exactly who gets told what and when, you do not have a communication protocol—you have a hope.

There is one honest reason organizations never do this: there is no forcing function. Incident simulations feel like busywork when nothing is actually broken, so they lose every scheduling contest to work that feels urgent. The forcing function that reliably works comes from the top. Have the board ask the question: "When did you last run an incident simulation? Walk me through one." Boards almost never ask it. The organizations where they do are, not coincidentally, the organizations that catch problems before they become disasters.

What Actually Works

Strip away the frameworks and the taxonomies and the argument reduces to a single, slightly deflating truth: the organizations that handle AI failure well are not the ones with the most advanced models or even the most sophisticated monitoring. They are the ones that planned for failure before it happened, wrote the runbook, ran the fire drill, and decided in daylight what the system should do at 2 A.M. When their system fails (and every system fails eventually), they do not panic. They activate the documented fallback, the application keeps working in a degraded but deliberate way, and they fix the root cause without a gun to their head.

Most organizations will have an AI-related incident within the next three years. The variable that decides whether that incident is a learning opportunity lasting a day or a crisis lasting a month is almost never the quality of the model. It is whether you built an operational discipline that has actually been tested, or one that merely exists on a wiki. You can find out which you have in about ninety seconds. Ask for three things about your highest-stakes system: the failure-mode inventory, the signed fallback sentence, and the date of the last drill. Three artifacts, three answers, no preparation. If anyone has to go looking, you have a wiki. It is paperwork and fire drills, and it is the only kind of work that holds at 2 A.M.

The Bottom Line

AI systems fail in ways that do not trip traditional alarms, and the failures that hurt most are the silent ones: the model working exactly as designed while the business degrades underneath it. Accuracy is not resilience: being right most of the time is worthless if the system has no way to catch, contain, or recover from the times it is wrong. Guard against all three failure types, not just the loud Type One crash you already monitor, because the silent Type Two and the treacherous Type Three are both more common and more damaging. Design fallback, escalation, and communication before deployment, as deliberate business decisions rather than 2 A.M. improvisations. And treat your operational risk framework as something to be rehearsed under stress, not documented and filed, because the difference between a framework that prevents disasters and one that merely describes them is whether it has ever been tested.

Field Kit

1. The failure-mode inventory (run before any AI system ships). For each system, force answers to: What if the model is confidently wrong about a *specific* type of situation? What if the input distribution shifts? What if the system's own outputs create a feedback loop that degrades it over time? What if volume spikes and latency climbs? What if the input stops arriving entirely? For each mode, write one sentence on what happens to the business—and one sentence on the designed response.

2. The three-type coverage check. Put your current monitoring in three columns. *Type One (infra):* uptime, latency, errors—likely covered. *Type Two (silent bad output):* baseline output distribution, continuous drift measurement, cohort/fairness monitoring—usually the gap. *Type Three (test-vs-production divergence):* production hold-out validation within a week of deploy, feature-distribution comparison, a tested rollback. Any empty cell is an exposure. Start filling Type Two first.

3. The fallback decision (one line per system). Complete this sentence before go-live: "When [system] is unavailable or degraded, the application will [serve yesterday's output / serve a safe default / degrade to fewer results / hold for human review / show a graceful error], and we are willing to run this way for [duration] before it becomes an incident." If no human has signed that sentence, you have no fallback.

4. The failover runbook (the part everyone skips). Explicit steps, tested in a non-production environment: what signals declare the system unhealthy; who gets notified and how; the exact command that activates fallback; how you confirm the AI is healthy again; the steps to roll back to normal; and the maximum time on fallback. If it has never been run as a drill, assume it does not work.

5. The quarterly incident simulation. Pick one failure mode per quarter—accuracy drop, latency spike, cost explosion, output-distribution shift, rising false positives. Walk the whole response live. Pass conditions: the on-call engineer reaches a "kill it or investigate" call within 2 minutes, the escalation decision is made within 5, internal communication completes within 15, and the team can name the external-communication steps within the hour. If a full walkthrough takes more than 30 minutes, the process is not ready—fix what broke and re-run next quarter.

6. The board question that forces all of the above. Give leadership one question to ask, every quarter: "When did we last run an incident simulation, and can someone walk me through it right now?" It is the cheapest forcing function that exists, and the organizations that ask it are the ones that catch problems before their customers do.

CHAPTER 14

Resilience, Risk Concentration & Predictive Governance

Part V · Operations: Monitoring, Failure & Resilience

The head of data science at an online home-goods retailer drew her model inventory on a whiteboard for me, and one box had eleven arrows pointing into it. It was the product ranking model, built three years earlier by two people to answer one narrow question: what to show on a category page. It worked. Conversion climbed. Search adopted it, then merchandising, then the email team, then the mobile home screen. By the time she drew that whiteboard, fifteen features depended on it and it shaped roughly four hundred thousand shopping sessions a day. Nobody planned this. There was no rollout committee, no adoption strategy, no architecture review that decided this model should sit at the center of the business. It happened the way weight accumulates on a strong branch: the model was working, and working models attract dependencies.

Then a third-party product-attribute vendor changed a schema. Not dramatically—one category field started arriving with a different encoding, announced in a release note nobody on her team was subscribed to. The model had never been trained for that case, so it started ranking differently, and the ranking was worse. Category-page conversion slid about four percent.

Here is the question that decides how bad this gets: how fast do you catch it? Monitored well, maybe two hours. Not monitored well, two days. Her team took thirty-one hours, and those thirty-one hours cost somewhere north of eight hundred thousand dollars in lost margin—not because the model was bad, but because the failing system was her most successful, most integrated, most depended-upon one. It was degrading the business precisely *because* it had been good enough to wire into everything.

This is one of the most reliable patterns behind AI failures in large organizations, and it is almost never said: your most successful model is also your biggest risk. Not your worst model. Your best one. The chapter that follows is about three linked truths that most enterprises learn the expensive way: that success concentrates risk, that surviving a crisis is a different discipline than performing well on a normal Tuesday, and that the organizations pulling ahead have stopped waiting for problems to announce themselves and started predicting them.

Here is the reframe this chapter is built on.

The myth: the models that put your business at risk are the weak ones, the ones that underperform or fail validation.

The reality: your most successful model is your biggest risk, because success is what wires a model into everything and raises the cost of it being wrong.

Success Is How You Build the Vulnerability

Walk the sequence again slowly, because the danger lives in how ordinary each step looks.

A model gets built to solve a real problem. It is carefully tested. Accuracy is good. You deploy it, and it works, and the business improves. So it becomes more important. A second team starts using the output, then a third. Now it is powering decisions across the organization. As adoption spreads, the model becomes more integrated: applications depend on it, business processes are built around it, people's workflows are optimized on top of it. And with every new dependency, the cost of the model being wrong goes up.

The moment you truly see this, you should feel uncomfortable. Your best model (the one that is most successful, most adopted, most integrated) is also the one whose failure would do the most damage. That is not a flaw in the model. It is a property of success itself.

Your CFO already understands this instinct, even if they have never applied it to machine learning. Put all your capital in one stock and you are exposed to that one company's collapse. Let all your revenue come from one customer and you are exposed to that customer walking away. Concentration creates vulnerability—every risk officer knows it in their bones. In finance it has a name and a whole discipline built around measuring and limiting it. In machine learning the identical principle applies and almost nobody names it: the more important a model becomes, the more of your risk is concentrated in that single system.

The treacherous part is that concentrated risk is invisible right up until it is catastrophic. The model keeps working. It keeps being accurate. Everything looks fine. Then the input distribution shifts, or the world changes, or an assumption the model was built on stops holding—and suddenly your most important model is also your broken model, breaking at the exact moment it is most deeply wired into your business.

Contrast the ranking model with a bounded one: a model used by one team, integrated into one feature. When the bounded model breaks, one feature gets worse. You turn it off, you lose that feature's value, and the business survives. When the concentrated model breaks, it breaks your most critical business process. Same failure, wildly different blast radius. The only variable that changed is how much you let the model matter.

Nobody Concentrates Risk on Purpose

Here is the myth: risk concentration is a mistake careless organizations make. The reality is that risk concentration is what happens to careful organizations that succeed. Almost no one deliberately concentrates risk. They stumble into it, one reasonable decision at a time.

The model was never *supposed* to be this critical. It was supposed to be one feature among many. But it was good, so it got adopted, and adoption is the thing that promotes a nice-to-have into a load-bearing wall. The monitoring was never *supposed* to be inadequate, but building a new model is fast and building monitoring infrastructure is slow, so the monitoring always lags the criticality. The retailer's ranking model was still watched the way it had been watched in year one, when it served one page: a weekly A/B dashboard, reviewed on Mondays. Fourteen new dependents had arrived since then and not one of them came with a new signal. The operational procedures were never *supposed* to be immature either, but nobody believed this model would become this important, so nobody invested in operational maturity to match the importance it eventually acquired.

Trace the dangerous pattern and it is almost boringly consistent: build a good model; it works, so it gets adopted; adoption outruns the rate at which monitoring infrastructure scales; the model becomes mission-critical before the operational maturity to support it is in place; the model breaks; the impact is enormous. The gap between "this model is now critical" and "we have the operations to run a critical model" is exactly the window in which concentrated risk turns into a crisis. Success opens that gap. Only deliberate effort closes it.

Managing Concentration Before It Manages You

You cannot make concentration risk disappear, but you can decide, on purpose and in advance, how much of it you are willing to carry. There are really five levers, and mature organizations pull some deliberate combination of them.

The first is the least fashionable: **limit criticality on purpose.** Not every model should be integrated into everything. Some models should stay bounded by design: "this model serves one product team and one feature; if it breaks, we turn it off and the feature degrades gracefully, but the business keeps running." This takes discipline, because the temptation is always to plug your best model into everything you have. Limiting criticality is choosing a smaller blast radius over a slightly better metric, and that is a genuinely hard trade to make when the model is performing beautifully.

The second is **redundancy for the models you cannot bound.** Some systems are going to be critical no matter what you do; for those you build a fallback that can carry the load when the primary fails: a simpler and more robust model, rule-based logic that covers the common cases, or a human decision process that can take over. Chapter 13 covered what makes a fallback real rather than filed. The concentration-specific point is narrower: the fallback is slower and less optimal than the primary, and that is exactly why it works. It converts a total failure into a degraded one, which is the whole game when a single model carries fifteen dependencies.

The third is the hardest because it demands spending before the pain is obvious: **invest in operational maturity ahead of criticality.** If you have a model trending toward becoming mission-critical, start building the monitoring, alerting, and incident response *now*: before it is critical, before you are trying to understand it for the first time in the middle of an incident. This is the investment that always loses the scheduling contest to whatever feels urgent, and it is the one that separates organizations that get surprised from organizations that do not.

The fourth is **spreading criticality across multiple models** rather than concentrating it in one hero. Instead of one recommender doing everything, run several doing different things, so that when one breaks the others still work and the system degrades gracefully instead of failing whole. It costs more to build; it buys down concentration.

The fifth ties the other four together: **monitor concentration explicitly, as its own metric.** Track which models carry the highest business impact. Track how that is changing over time. Track which models are wired into the most systems. When a model starts trending toward high concentration, treat that trend itself as the signal: the trigger to invest in operational maturity and risk management before, not after, the model becomes indispensable. Most organizations monitor accuracy obsessively and monitor concentration not at all, which is how they end up genuinely surprised that their most reliable model became their biggest crisis.

Robust Is Not the Same as Resilient

Everything so far assumes a harder question sitting underneath it: when a critical system does break (and eventually one will), does it fail catastrophically, or does it bend and survive? That question exposes a distinction most enterprises never make explicitly, and it is the difference between robustness and resilience.

Your system needs to be robust. It needs to work under normal conditions: tests passing, models performing, pipelines running. But robustness is not enough, because robustness is only ever about handling *expected* conditions well. Robustness is about being strong: the model is accurate on test data, it handles normal input ranges, it works when deployed. Resilience is about bouncing back: when something breaks, the system does not simply fail; it degrades gracefully, falls back, responds, recovers.

A robust AI system handles the normal cases beautifully and fails catastrophically the moment something unexpected arrives. A resilient system might handle the normal cases slightly less optimally (there is real overhead in resilience), but when the crisis comes, it survives. And this is where most enterprises are exposed, because most enterprises optimize entirely for robustness. You test the model, validate it, monitor it, deploy it, and it works. Then something unanticipated happens: the distribution shifts in a way you did not model, a dependency breaks, a fraud pattern mutates, a regulator announces new requirements overnight. The system was robust against everything you tested for and fragile against everything you did not. That is the anatomy of nearly every "how did such a well-built system fail so badly" story you have ever read.

Resilience requires designing for failure: not “expect specific things to fail” but “assume things will fail in ways we did not predict, and design to survive anyway.” Chapter 13 covered most of the machinery that assumption cashes out as: **graceful degradation**, so a struggling system steps down instead of crashing or spewing bad output; **fallback systems**, so critical decisions have somewhere to go when the AI cannot handle them; and **explicit crisis decision authority**, so a named person can pull or restrict a model at 2 A.M. without convening anyone. Assume all three are built. Two more capabilities matter specifically for the concentrated systems this chapter is about, and they are the two that organizations most reliably skip.

Crisis-mode observability starts from an uncomfortable fact: most monitoring is designed for normal operation, and under real stress the monitoring is often the second thing to fail. A regional health insurer learned this with a claims-triage model whose dashboards read from the same feature store the model did. When the feature store began serving stale values, the model triaged on stale features, and the dashboards, drawing from that same source, froze at the last good numbers. Every panel stayed green for six hours. The team found out from a claims supervisor who noticed her queue had stopped looking like her queue. The fix was not a better dashboard. It was a deliberately stupid one: five numbers (request volume, response volume, output distribution, error rate, and time since the last fresh input) sampled straight off the serving path, written to storage that shares nothing with the production pipeline, and rendered on a single page. It tells you far less than the real monitoring. It has the singular virtue of still working when the real monitoring does not. The question to ask of every concentrated system is short and rarely asked: which of my monitors depend on the thing they are monitoring?

Pre-planned recovery means the recovery is designed and tested before the crisis rather than invented during it, and the useful version of it is quantified. If the model fails, we switch to the fallback: it takes ninety seconds, it happens automatically below a confidence threshold and requires a human above it, and here are the three signals that confirm the fallback is actually carrying the load rather than silently doing nothing. Then comes the half that almost nobody writes down—the trip back. What has to be true before you return the primary to service? Who makes that call? And what do you do about the decisions made while you were degraded: the recommendations served, the claims triaged, the prices quoted, the applications routed. Most recovery plans stop at “switch to fallback” and leave the return and the cleanup to improvisation, which is how a two-hour outage becomes a two-week reconciliation. The higher the concentration, the larger the pile of decisions you have to go back and account for.

Why Resilience Loses Every Budget Fight

Resilience is expensive, and worse, it is expensive in exactly the way organizations hate. You are building fallbacks that may never be used. You are monitoring in ways that look like pure overhead. You are designing for a failure you would rather believe will never come. The cost is highest in normal times (fallbacks consume resources, extra monitoring adds latency, crisis-mode decision authority requires training and drills), and the payoff only ever arrives in the crisis, when the resilient system stays up and the robust one falls over.

So most enterprises skip it, for reasons that are honest and a little embarrassing. It is invisible in normal times: robustness is visible because the system works, while resilience is invisible because you never notice the fallback you never needed. It requires thinking about failure, and organizations are psychologically wired to think about success. And it requires ongoing discipline: testing fallbacks, running drills, maintaining decision authority, none of which is technically interesting and all of which loses to whatever is on fire that week.

The work that protects you is boring, which is precisely why so few organizations do it. What separates the ones that do is not effort but a design choice: they stop trying to optimize robustness and resilience with the same knob, because the two pull against each other. They optimize robustness for the normal cases through good design, testing, and monitoring, and they optimize resilience for the crisis cases through fallbacks, simple decision authority, and pre-planned recovery. Different targets, different design choices, both on purpose.

Most enterprises pay the real cost of skipping resilience exactly once: one catastrophic failure, and then the incident itself is the cheap part. The expensive part is the lost trust, the regulatory attention, the months spent rebuilding. After that, they build resilience. The smart ones build it before they pay.

From Monitoring the Past to Predicting the Future

There is a moment when an organization shifts from reactive to predictive in any domain, and it almost never happens because someone circulated a paper. It happens because reactive stopped working.

You can run a reactive compliance program as long as audits stay infrequent and deployments stay controlled. You can run reactive model risk management as long as failures stay local. But at scale (with continuous deployment, with mission-critical systems depending on AI), reactive becomes unworkable. You are always a quarter behind. Your governance team spends all its time investigating problems that already happened instead of preventing the ones that have not. The leading enterprises have already moved past this. They are building predictive governance, using metrics that tell them not just what happened but what is likely to happen next.

The distinction is sharper than it sounds. Reactive governance answers a single question: *is there a problem?* You run audits, check compliance, measure performance, investigate incidents. All necessary, all permanently one step behind reality. Predictive governance answers a different question: *where is a problem likely to develop?* You watch metrics that lead rather than lag: drift, volatility, distribution change in real time; decision velocity; the places where an assumption is most likely to fail before any system actually fails.

This works because most problems in AI governance do not arrive suddenly. They develop. Assumptions drift, data distributions shift, edge cases multiply, team knowledge fades. Watch the right signals and you see it coming. The reactive metrics everyone already tracks (audit pass/fail, incident count, validation-set accuracy, checklist completion) tell you what already happened, and by the time you have the audit result the audit is old. The predictive metrics tell you what is starting to happen.

Distribution stability asks not whether the input data changed but how fast and in what direction it is changing week over week. High volatility in the input distribution predicts drift in model behavior *before* it shows up in performance metrics: if your inputs are shifting rapidly, the relationships your model learned are going stale, and you can know that while accuracy still looks fine. The home-goods retailer's vendor schema change would have registered here as a step change in the distribution of a single categorical field, hours before it registered as a conversion drop. **Assumption coherence** is the metric form of Chapter 7's assumption register: do the assumptions you deployed under still hold, checked continuously rather than at review time? When reality starts diverging from what is written in the register, governance failures are already in flight. **Decision velocity**, Chapter 3's standing metric, belongs on this list for the same reason: the lag between finding a governance issue and deciding about it predicts whether issues will start accumulating faster than you resolve them. **Explanation consistency** asks whether your team gives you the same account of why a model behaves as it does, or a different one each time. Low consistency is a quiet signal that real understanding is fading and process is calcifying without comprehension. And **engagement depth** measures how substantively your best engineers engage with risk, as opposed to performing compliance; low engagement is a leading indicator of governance that will collapse the first time it is genuinely stressed.

Governance as Early Warning, Not Enforcement

Shift to predictive governance and your governance team's role changes underneath them. They stop being only auditors and enforcers and become risk interpreters. The data science and product teams are constantly generating signals about where problems might develop; the governance team's job is to read those signals before they become incidents. That demands different tools and different access: real-time, stream-level visibility into data distributions, model behavior, and decision velocity rather than reports after the fact. It means assumptions checked continuously rather than at review time, velocity metrics on the governance process itself, and behavioral observability that shows where users get different outputs, where confidence is high but accuracy is low, where edge cases are surfacing. All of it is achievable with today's technology. Most enterprises have not built it, for the simple reason that reactive mode does not require it.

It also changes the relationship with your technical teams, and this is the part that compounds. When governance is reactive, teams experience it as enforcement arriving from outside, and they get defensive. When governance is predictive, teams experience it as early warning (someone helping them see problems before those problems become failures), and they share information more openly, which makes the predictions better, which builds more trust. For your board, predictive governance offers the one thing reactive governance structurally cannot: genuine foresight. Instead of "an incident occurred," the report becomes "we are seeing these risk signals emerge, and here is what we are doing before it becomes a problem." That is the whole distance between risk mitigation and risk management, and it is also what lets you move fast. The enterprises moving fastest on AI are not moving blind. They are moving with real-time visibility into their own risk, which is exactly what makes the speed defensible.

What Actually Works

Put the three ideas together and they collapse into one discipline. Your best model is your biggest concentration of risk, so you decide on purpose how much criticality it is allowed to carry. Robustness will get you through a normal day but only resilience gets you through a crisis, so you design for the failure you hope never comes. And waiting for an audit to tell you what already broke is, in a world of continuous deployment, waiting too long, so you build the metrics that let you see the next problem developing.

The organizations that handle this well do not celebrate a successful model and move on. They look at a model working perfectly and ask the uncomfortable question (*how could this break, and what would the impact be?*) and then they invest before the urgency is obvious, while there is still time to be deliberate. Concretely: saying no to the fifteenth integration when the existing fourteen already exceed what your operations can carry; funding an observability path that shares nothing with the pipeline it watches; writing down the return trip as well as the failover; and putting concentration and distribution stability on the same dashboard as accuracy, where someone has to look at them every week.

The organizations that handle it badly are the ones that will, three years from now, be genuinely astonished that their most reliable system became their most expensive crisis. The success was always real. The crisis was always avoidable. The only difference is whether anyone was thinking about concentration, resilience, and prediction from the start.

The Bottom Line

Your most successful model is your biggest concentrated risk, because success is exactly what wires a model into everything and raises the cost of it being wrong. Almost no one concentrates risk on purpose; they stumble into it one reasonable decision at a time while monitoring and operational maturity lag behind criticality. Robustness handles the conditions you tested for and fails catastrophically on the ones you did not; resilience is the separate, expensive discipline of degrading gracefully and recovering when the unexpected arrives. The two pieces most often missing are observability that survives the stress it is supposed to measure and a recovery path that covers the trip back, not just the trip out. Reactive governance can only ever tell you a problem already happened, so the enterprises pulling ahead are shifting to predictive metrics (distribution stability, assumption coherence, decision velocity) that reveal where the next failure is developing before it lands. Thirty-one hours of a degraded ranking model cost one retailer eight hundred thousand dollars, and every hour of it was avoidable by someone watching the right box on the whiteboard. The crisis is always avoidable. It just requires thinking about concentration, resilience, and prediction from the start rather than celebrating your best model and being surprised when it breaks.

Field Kit

1. The concentration map (one page, reviewed quarterly). List every model in production. For each, record its business impact if it fails (low/medium/high), how many teams and features depend on it, and the direction of that dependency count over the last two quarters (rising/flat/falling). Any model that is high-impact *and* rising is a concentration signal—treat the trend itself as the trigger to invest in operational maturity now, before it becomes indispensable.

2. The criticality decision (force it before the next integration). Before wiring your best model into one more system, answer in writing: "If this model fails, what breaks, and can we turn it off and degrade gracefully—or does it take down a critical business process?" If the answer is the second, you are choosing to concentrate risk. Choose it deliberately, or bound the model instead.

3. The resilience checklist for critical systems. For each high-impact model, confirm five things exist *and have been rehearsed*, not merely documented. Three come straight off Chapter 13's runbook—graceful degradation, a defined fallback, and named crisis decision authority—and the only question here is whether your highest-concentration models are actually on that runbook. The other two are this chapter's: crisis-mode observability, meaning a minimal signal set that shares no infrastructure with the pipeline it watches, and a timed recovery path that covers the return to primary and the cleanup of decisions made while degraded. Any item that is documented but never rehearsed should be assumed not to work.

4. The robustness-vs-resilience audit (one question per system). Ask: "What have we optimized this system for—normal cases, or crisis?" If every answer is "normal cases," you have a portfolio that performs well on a good day and has no plan for a bad one. Pick your highest-stakes system and close that gap first.

5. The predictive-metric starter (pick one, this quarter). You will not shift from reactive to predictive overnight. Choose one leading indicator for your highest-risk system—distribution stability, assumption coherence, or decision velocity—and get good at reading it before it becomes a crisis. Track it on the same dashboard as accuracy. Then add the next one.

6. The governance foresight question (for the board). Replace "what incidents did we have?" with "what risk signals are we seeing emerge, and what are we doing before they become incidents?" A team that can answer it has predictive visibility; a team that cannot is still governing the past. Asking it every quarter is the cheapest forcing function for building the visibility in the first place.

PART VI

The Economics of Enterprise AI

CHAPTER 15

Why AI Budgets Blow Up: Cost & Unit Economics

Part VI · The Economics of Enterprise AI

A VP of Engineering told me about her AI project. The budget was $2 million. The actual spend was $6.8 million. When I asked what happened, she said: "We didn't understand what we didn't understand."

That is a good line, and it is also a diagnosis—though not the one she thought she was giving. She heard herself confessing a failure of planning, as if better scoping in month one would have saved her. It wouldn't have. She budgeted for an AI project using a framework built for a completely different kind of work, and the framework did exactly what it was designed to do, right up until it fell apart.

Here is the reframe this chapter is built on.

The myth: AI budgets blow up because the first estimate was too optimistic, and tighter scoping in month one would have kept the project on plan.

The reality: AI cost is not something you forecast, it is something you control, with spending caps, named owners, decision gates, and a cost per unit of value.

The organizations that blow through AI budgets are almost never incompetent. They are running a forecast where they needed a control system, and they don't notice the mismatch until six months in, when the budget is half-spent and the project is twenty percent complete. What follows is not a model that predicts AI costs accurately, because there is no such model. It is the discipline that prevents the blow-up: a way of budgeting, deciding, and governing that assumes you will be wrong and gives you room to manage it when you are.

The Framework Was Built for a Different Kind of Work

Traditional software budgeting rests on four assumptions: you know what you're building, you know roughly how long it will take, you know what "done" looks like, and you know what success means. AI projects violate all four.

You think you're building a recommendation system. Once you start, you discover your data quality isn't good enough, and you spend six weeks fixing data. Then you realize your features don't predict what actually matters, and you spend eight weeks rebuilding them. Then the model works, but it produces biased outputs, and you spend another month on fairness. None of these were unknown risks. They were surprises only in the sense that you didn't know *which* of them would be the blocking issue until you hit it.

The fundamental problem is that AI projects are exploration work disguised as execution work. In traditional software you explore during design (architecture reviews, specs, planning) and then you build, and the build is well-understood because you have precedent and patterns and people who know how long things take. In AI the exploration never stops. You're building the model and learning what it can do at the same time, testing it and discovering what it fails at, deploying it and discovering what real-world performance looks like. Each phase reveals information that changes the phase that comes next.

So you budget for the plan you made in month one, and month three reveals that the plan was forty percent wrong. Do you stop and replan? Most organizations don't; they replan and continue. Now you're $800,000 into a $2 million budget, with $1.2 million left to finish work that, based on what you now know, actually costs $1.8 million.

Where the Costs Actually Hide

The budget doesn't blow up because AI is hard. It blows up because there are whole categories of cost that traditional frameworks fold into "engineering overhead," and then one of them explodes and there's no buffer. There are five worth naming, because you can only budget for what you're willing to look at.

The first is the **data tax**. In traditional software the data is usually already there; in AI you often have to acquire, clean, label, and prepare it, and a team can burn forty percent of the project doing so. One organization budgeted $200,000 for data prep and spent $800,000: they discovered their labels were inconsistent and had to relabel 100,000 examples, discovered they lacked geographic variation and had to acquire new data, discovered a feature time-window was wrong and had to rebuild the dataset. That is not incompetence. It is the ordinary cost of working with data you don't fully understand up front. And you don't understand it until you build the first model and see what's missing.

The second is the **iteration tax**. You build a model and it hits 87% accuracy. Is that good? You don't know without a baseline, and you don't know whether an 87% model that happens to be biased against a particular group meets your requirements until you measure it. So you iterate (different architectures, feature sets, training approaches), and a team might run thirty or forty iterations before landing on something both technically sound and acceptable to the business. Most organizations budget for "build the model," not for "understand the model well enough to deploy it," which is the part that actually costs money.

The third is **validation and governance overhead**. Before you deploy, you have to test edge cases, check for bias, handle distribution shift, and document failure modes. A data scientist can easily spend thirty percent of their time building and seventy percent validating and preparing for review. Most organizations assume "validate" is a 10-to-20% tax on top of "build." In practice it is frequently 50-to-70%, and it compounds when the governance structure is unclear—which, on a first AI project, it always is.

The fourth is **integration and infrastructure**. The model doesn't live in isolation; it lives inside your product, which means serving infrastructure, monitoring, logging, feedback loops, retraining pipelines, and fallback systems. This is often thirty to fifty percent of total project cost, almost never budgeted separately. It gets rolled into "platform overhead," and then halfway through the project it resurfaces as an entire second project that was never in the plan.

The fifth is the **expectation tax**. You deliver a working model, and the business stakeholders have ideas. What if we changed the optimization target? Added this constraint? Included this data source? Each is reasonable; each is a couple of weeks of work; three stakeholders with two ideas each is six iterations that were never in the spec. This one is nobody's fault, so nobody budgets for it: the normal cost of collaboration, unpriced.

Watch how these interact and the failure mode becomes obvious. The $700,000 of contingency you set against a $2 million project gets eaten by the data rework in month four, the serving rewrite in month six, and a stakeholder's new metric by month eight—until the contingency is gone, the data team is burned out, and you have $500,000 left to finish $800,000 of work. The project didn't fail because the AI was hard. It failed because the framework assumed these categories were fixed and independent, when they are variable and interlocked.

The Math That Actually Works

Budget correctly and the numbers move around in a way that looks strange the first time you see it. Here is what a real $2 million AI project can look like. A **discovery phase**: $300,000 over two to three months, whose output is not a production model but clarity on what the production model will require. A **build phase**: $800,000 over four to five months. **Governance and validation** as its own line: $300,000 for deep validation, bias testing, failure-mode analysis, and documentation. **Infrastructure and serving** as a separate project: $400,000. And genuine **iteration and contingency**: $200,000, ongoing, for the stakeholder changes and performance work you already know are coming.

Notice three things. The "model" (the part everyone pictures when they picture an AI project) is maybe forty percent of the cost; everything else is the work of making it real and acceptable. Governance is a line item with an owner, not a tax hidden inside someone else's number. And the contingency is earmarked for things you expect to happen, not for freak events. This framework isn't better because it forecasts accurately—it doesn't. It is better because it accounts for the actual sources of cost. Your next AI project will overrun somewhere; the only question is whether you budgeted for that or pretended it wouldn't happen.

The More Expensive Failure Is Not Stopping

A correct budget protects you from one kind of failure. There is a second kind it does nothing for, and it is the more expensive one.

There is a moment in most AI programs where someone needs to ask whether to keep going: not whether it is theoretically possible, not whether the problem is interesting, not whether the vendor says it will work, but whether the organization should keep spending money on it. And organizations ask this question late: after they've overspent, after they've built organizational muscle around the program, after stopping has become politically difficult. By then the conversation has already been decided by sunk costs.

"We've already invested $5 million. We can't just walk away." "We've been at this eighteen months. We need to see it through." "The executive sponsor is committed." These are the same argument wearing three coats: we already paid, we can't get it back, so we might as well continue. The logic feels responsible. It is exactly backward.

I watched this play out at an organization that had committed to an enterprise-wide machine learning platform. The rationale was sound: many models, one system to manage them all, a vendor who claimed a clean fit with their stack, an enthusiastic internal team. Eighteen months in, the integration was harder than expected, the maintenance overhead larger than planned, the payoff hazy. More to the point, the business had moved on; different problems now seemed more urgent. The team kept building anyway, because they'd already spent enough that leadership hesitated to pull the plug. A year later the project was canceled. The system was never deployed. All of it—gone.

That is not a vendor failure or a technical failure. It is a failure to ask the right question at the right time, and the right question is almost embarrassingly simple: *if we could start over today, knowing everything we now know, would we build this?* If the answer is no, you probably shouldn't keep building it—no matter what the counter reads.

What makes sunk costs so dangerous is that the amount you've already spent tells you nothing about how much more you should spend. The only thing that governs the next dollar is future expected value. The $5 million is gone whether you continue or quit, which makes it irrelevant to the decision in front of you. The question is never "can we afford to stop?" It is "can we afford to keep going?"

The organizations that handle this well treat AI programs like venture capital investments: high risk, frequent decision points, explicit go/no-go criteria. Every program gets business milestones, not just technical ones. At month six: do we now know whether the business case holds? At month twelve: does the architecture actually work? At month eighteen: do we have evidence the model improves the process? And at each gate the question is not "did we hit our technical targets?" but "has anything changed about what makes this valuable, and would we start this today?" Few organizations build these gates, so they drift; the sunk costs mount, the decision gets harder, and the project becomes its own justification.

Something usually does go wrong, not from incompetence but because the future is hard to predict, and what looked like a good use of capital eighteen months ago may not be one today. The failure is not adapting when the uncertainty resolves, which makes the real prerequisite cultural: an organization where saying "this isn't working anymore" is a normal part of learning, not a career-ending admission. Naming a losing bet early is the whole skill.

From Projects to Production: The Costs That Never Stop

Everything so far has been about projects, things with a beginning and an end. The nastiest AI costs don't have an end. They start the day you succeed.

Most organizations ask the wrong question about ongoing AI cost. They ask "how much will this initiative cost?" when they should ask "how do we keep this cost from spiraling?" Forecast versus control again. It matters because AI cost doesn't behave like traditional IT cost. With infrastructure you pay for capacity, and the cost is relatively predictable; you plan for growth and you scale. With AI you pay for inference, and inference scales with usage. As the model gets more popular and lands in more applications, the cost climbs on its own, and because a single inference is cheap, nobody feels any urgency to optimize it.

The progression is always the same. Someone builds a model that works and it gets popular. Inference costs $0.02 a request, which is nothing—except requests go from 10,000 a day to a million a day, and now you're spending $20,000 a day on a model nobody planned to spend $20,000 a day on. It just happened. So you face a choice: shut it down, or keep paying. Because it's popular and it works, you keep paying, subsidizing that usage out of budget meant for new models. Repeat across a year and you've deployed seven models; three are eating sixty percent of your inference budget, four are barely used but still running, and you're carrying several million dollars of unbudgeted spend, scrambling to find where it came from.

The usual reactions ("we need to be more careful about which models we deploy," "we need better forecasting") both miss, because the problem is neither selection nor forecasting. It is that you have a cost-*tracking* structure and not a cost-*governance* structure. You measure what you spent. You don't control how much you spend.

Cost governance does not mean "spend less." It means "make spending decisions consciously," and in practice it is five habits. **Every model has a cost budget, not a**

forecast: “this model can spend up to $50,000 a month,” and if it exceeds that, it gets optimized or gets an explicit, deliberate increase. **Every model has an owner who cares about cost**: not the data scientist who built it (the data scientist cares about accuracy) but the product manager or business owner of the process it serves. **You review model costs monthly**, asking why a model consuming forty percent of the inference budget isn’t producing forty percent of the value. **You have a retirement process**, because otherwise models accumulate into a graveyard nobody will switch off in case someone, someday, needs it. And **you look for economies of scale** (batching, smaller models, cheaper regions, caching), optimizations that only become visible once you have cost ownership to begin with.

There is an uncomfortable part, and it is the part organizations most want to avoid. Real cost governance forces conscious trade-offs. With a limited inference budget you sometimes have to choose between a very accurate model run infrequently and a less accurate model run often, and someone has to say, in a meeting, “yes, we are fine with 85% accuracy if it costs half as much.” Most organizations would rather say “let’s optimize both” or “we’ll figure it out later.” And that avoidance is precisely what manufactures runaway cost. The ones that hold the conversation end up with inference costs twenty to thirty percent lower, because they keep asking “do we still need to run this?” and “can we run it cheaper?” instead of assuming it will all work out. They never open the quarterly bill surprised. That is what control feels like.

Unit Economics: The Question Your Board Is About to Ask

Zoom out from any single model to the whole enterprise and a bigger, more dangerous gap appears. Most organizations know their annual AI budget: they’re spending $50 million, or $200 million; pick a number. What they cannot tell you is anything *per*: per model, per use case, per business outcome. They don’t know whether a given model costs $10,000 or $10 million, whether training or inference or data prep is the dominant expense, whether consolidating would halve their spend. They’ve spent $50 million and have no idea whether they got value for it.

The structure of AI spending is what makes it invisible. The model is built by data science; the compute cost lives in the cloud bill; the data prep sits in the analytics budget; monitoring is buried in ops; maintenance is smeared across teams. So when you look at your P&L you see a cloud bill, an analytics line, an ops budget, and engineering headcount. You do not see “AI.” The common fix (create an “AI budget” and track it by project) gives you visibility into *planned* spending and does nothing for *actual* spending. Each cost center optimizes its own piece, and nobody optimizes for the only question that matters: whether a given model is worth what you're paying for it.

Unit economics is the answer, a simple idea borrowed from businesses that have to survive: how much does it cost to deliver one unit of value? In e-commerce, $50 to acquire a customer against $30 of margin means you're losing money; $5 against $30 means you're winning. Translated to AI, five numbers do most of the work.

Cost per model per year: training, inference, monitoring, maintenance, retraining, and fixes, divided by the years you expect to run it. A single model at $2 million a year should shock you into asking whether it's worth $2 million. **Cost per inference**, for high-volume models: some cost $0.001, some cost $1, and across a billion inferences a year the gap between $0.001 and $0.01 is $9 million. **Cost per business outcome**, the hard one, which ties the model to revenue, risk reduction, or savings (the fraud model against fraud prevented, the conversion model against incremental revenue), including what it costs when the model is wrong. **Cost per model update**: a model retrained weekly at $50,000 a cycle is $2.6 million a year in retraining alone, and you should know whether the gain over monthly is worth three-quarters of that. And **the cost of model-induced problems**, which almost nobody measures: when the model breaks, what is the customer impact, the engineering time, the time-to-detection? Plenty of organizations run models that could cost millions if they failed and have never calculated the number.

Once you know these numbers, the decisions make themselves. You find seventeen models doing similar things at $500,000 each that could consolidate into two at $200,000 each. You find model A at $2 million a year and model B at $500,000 delivering similar value, and you retire A. You find a model retrained weekly when monthly would give nearly the same accuracy at a quarter of the cost. None of these questions get asked without the numbers. And, most importantly, neither does the one that matters for the *next* model someone proposes: is the cost reasonable against the value? You decide from first principles instead of from enthusiasm for the technology.

Here is where it gets hard, and where most attempts die: the costs are owned by different teams. Data science owns modeling, infrastructure owns compute, ops owns monitoring, product owns the retraining decision. To compute unit economics you have to break the silos and make each team report its cost by model, which is annoying, and you will hear "that's not how our cloud billing works." The honest truth is that you're not measuring it because it's hard, not because it's impossible. And in the meantime you are making billions of dollars of decisions without the information you need.

Building it is real work (cost attribution, centralized tracking, monthly reporting, governance tied to the numbers) with a real cost in engineering time and discipline. So is the payoff: you stop paying for models that aren't worth it, consolidate the overlapping ones, and invest where the economics actually work.

The reason this matters lands in a single room. Your board asks: "What are we getting for our $200 million AI spend?" With unit economics you can say: "This model generates $50 million in value a year and costs $5 million to run. This one costs $3 million and prevents $30 million in fraud risk. This one costs $500,000 and generates $2 million in value, so we're retiring it." Without unit economics your answer is: "We're investing in our AI future. We're building capabilities. It's hard to measure." The first answer is leadership. The second is theater.

What Actually Works

Strip away the frameworks and what's left is a single, slightly deflating idea: AI costs blow up because organizations treat cost as something to forecast, when it is something to control. None of the fix is technically interesting: no architecture puzzle, no machine-learning breakthrough, just attribution, budgets, reviews, retirement processes, and the willingness to say out loud whether an 85% model that costs half as much is good enough. That is exactly why so few organizations do it, and why the ones that do stop being surprised by their own bills. They know what they're spending, because they chose to spend it.

The Bottom Line

AI cost overruns are not caused by AI being expensive; they are caused by budgeting frameworks that forecast when they should control. Traditional project budgeting assumes you know what you're building, how long it takes, and what done looks like. AI violates all three, hiding real money in data prep, iteration, validation, infrastructure, and stakeholder churn that never appear as line items. The same failure recurs at every altitude: projects overrun for lack of a buffer, programs zombie forward because sunk costs replace expected-value thinking, production inference spirals because usage scales cost with no budget cap, and enterprise spend becomes unaccountable because no one can trace a dollar to a model. The fix is unglamorous and consistent: spending caps with owners, decision gates that ask "would we start this today?", and unit economics that put a cost-per-value number on every model. Do that and the costs stop taking care of themselves—which is the point, because taking care of themselves is exactly what they were doing when they blew up.

Field Kit

1. The five-tax budget check (before you fund any AI project). Force a separate, explicit number for each hidden cost category, not a lump "engineering" line: the data tax (assume a 40% buffer and that you'll rework data), the iteration tax (budget 30–40 model iterations, not "build the model"), validation and governance (assume 50–70% overhead with a named owner, not a 10% tax), integration and infrastructure (a separate project at 30–50% of total if you're building new), and the expectation tax (1–2 stakeholder-driven iterations you know are coming). If any of the five is zero or buried inside another line, that's where the overrun will come from.

2. The reallocated-budget template. For a project of size *X*, split it into five funded phases instead of one build: Discovery (~15%, output is clarity not a model), Build (~40%), Governance & Validation (~15%, its own owner and timeline), Infrastructure & Serving (~20%, a separate project), and Iteration & Contingency (~10%, earmarked for things you expect). If "the model" is more than half your budget, you have not budgeted for making it real.

3. The go/no-go gate questions (bring these to every milestone review). Ask, in order: If we could start this project today, knowing what we now know, would we still build it? If no, what would we do instead? How much more should we spend before we decide? Forbid "we've already spent $X" as an input; the only valid input is future expected value. Set the gates—month 6 (business case), month 12 (architecture), month 18 (evidence of process improvement)—before the project starts.

4. The production cost-governance card (one per deployed model). Fill in before go-live: monthly cost budget (a cap, not a forecast); named cost owner who is *not* the data scientist; review cadence (monthly minimum); the action when the cap is hit (optimize / raise budget deliberately / retire); and the retirement trigger. A model without a signed card is a model quietly authorized to spend without limit.

5. The unit-economics scorecard (per model, reviewed quarterly). Five numbers, sourced across teams: cost per model per year, cost per inference, cost per business outcome, cost per model update, and the estimated cost of a model-induced failure. The scorecard's job is to surface the three killer questions: Which models overlap and could consolidate? Which cost far more than a similar-value alternative? Which are retrained more often than the accuracy gain justifies?

6. The board answer test. Draft the answer to "What are we getting for our AI spend?" as if the board asked today. A passing answer names specific models with cost-versus-value figures ("$5M to run, $50M in value; $3M to run, $30M in fraud prevented; retiring the $500K model that returns $2M"). If the honest draft is "we're building capabilities, it's hard to measure," you have theater, not leadership—and the gap between the two drafts is your unit-economics to-do list.

CHAPTER 16

Talent & Capability: Stop Hiring Unicorns

Part VI · The Economics of Enterprise AI

"Looking for an AI/ML Leader with 10+ years of machine learning experience, deep understanding of governance, strong executive presence, ability to translate technical concepts for the board, a proven track record scaling data science teams, and experience in regulated industries."

You have seen that posting. You may have written it. Read it again slowly, because it is a fantasy. It describes someone world-class at machine learning research, world-class at organizational design and governance, and world-class at executive communication, all at once, in one body, for one salary. It describes a unicorn. And the search for that unicorn is sinking your AI program.

The person who is genuinely world-class at machine learning research does not want to spend 40 percent of their week in governance meetings. The person who is genuinely world-class at organizational design does not have a publication record at top ML conferences. The person who can hold a room full of board members does not think in loss functions. These are not the same person. They are almost never the same person. And the market you are actually hiring from does not contain the exception you're praying for.

Here is the reframe this chapter is built on.

The myth: AI capability is something you acquire by hiring the right individual, one leader who is world-class at models, governance, and the boardroom at once.

The reality: you cannot acquire AI capability by finding the right individual; you build it into the structure of your organization, and that is truest of all in governance.

That's true at the top, where you're hiring a leader. It's true in the middle, where a single engineer is drowning. And it's truest of all in governance, where the entire industry is trying to hire its way to something that can only be grown. The mistake repeats at every altitude, and it always looks the same: you go shopping for a person when you should be building a system. This chapter is about stopping.

Three Functions, Not One Person

Start by separating what you're actually asking for. Underneath that impossible job posting are three distinct functions that genuinely need to exist. They are real. What's unreal is expecting one hire to hold all three.

The Model Builder understands how to train, fine-tune, or integrate foundation models. They know the technical trade-offs, they can debug strange model behavior, and they can estimate compute and inference costs before the bill lands. This is a technologist. They might hold a PhD, or they might have five years of production experience, or they might have started last year and simply be brilliant. The credential doesn't matter. The trait that matters: they know AI systems can be built, and they know how to actually build them.

The Program Governance Architect understands how AI systems fail inside organizations. They know how to build accountability structures, which risks matter and which are for show, how to translate technical capability into business constraint, how to design an approval process that actually functions instead of one that merely exists. This person probably can't build a model, and doesn't need to. They understand organizational design. They've worked in regulated environments. They've watched programs fail and they know exactly why. They come from operating experience, not from research.

The Executive Translator can stand in front of a board or a CFO and articulate what's possible, what's risky, and what's strategic about AI. They may not understand backpropagation, and they don't care to. What they understand is business impact, board dynamics, shareholder communication, and how decisions actually get made in a C-suite. Their gift is turning "the model drifted" into "our margin exposure is X, and here's what I recommend."

What organizations do (what you may be about to do) is try to hire one person for all three. The result is predictable: you hire someone adequate at all three and excellent at none. You've paid a premium for a compromise.

Why the Unicorn Hire Fails

The unicorn feels safer than the alternative, and that's exactly why it's so seductive. One person owns AI. One P&L, one budget, one accountability line. Your board likes it. Your CFO likes it. It's clean. It fails for three reasons, and they're worth naming because each one shows up eighteen months later disguised as a surprise.

First, you've built a single point of failure. If the person leaves or underperforms, the whole program slows or stops. You didn't build organizational capability; you rented individual capability. And individuals leave.

Second, you've manufactured role confusion. One human doing three jobs will prioritize the one they enjoy and are good at, and starve the rest. Almost always, the one they starve is governance. This isn't a character flaw: it's arithmetic. They're spending 70 percent of their time building models and 20 percent talking to executives, which leaves 10 percent for governance. Then, a year and a half in, you're shocked to discover you have essentially no governance structure. You shouldn't be shocked. You designed the incentive that produced it.

Third, you hired against a market that doesn't exist. Because no one is world-class at all three, you settled for someone medium at all three. Six months later you learn they're not technically deep enough for your hardest problems, or not organizationally sophisticated enough for your real governance needs, or not credible enough in front of the board. You filled the role and solved none of the underlying problems. The seat is warm and the work isn't happening.

What Actually Works: A Structure, Not a Savior

The organizations whose AI programs are actually running (generating value, not lurching from crisis to crisis) don't have a better unicorn. They have a different shape. It's typically three roles working in explicit, deliberate collaboration, plus an executive who keeps them honest.

The **Chief AI Officer** or AI Program Lead is the integration point. They own the P&L, the roadmap, the budget. But they don't pretend to be the best in the room at everything. In practice they're something like 40 to 50 percent governance architect, 30 to 40 percent operator, and 20 percent translator. And they are entirely comfortable not being the sharpest pure-ML mind in the building. That comfort is a feature, not a gap.

The **Technical Leader** (a VP of Data Science, a Head of Applied AI, a Principal Architect, the title varies) owns the actual execution. Model builds, infrastructure, performance. They are not trying to become a board-level communicator or an organizational designer. They are trying to make the systems work, and that is a full and honorable job.

The **Governance and Risk Lead** owns the structures, processes, and accountability mechanisms. They might report into the CAIO, or they might sit in Risk, Compliance, or Legal alongside the CRO and the GC. Wherever they sit, they are explicitly responsible for *designing* governance, not merely for "making sure we follow the rules," which is a clerk's job, but for building the machinery that makes good decisions repeatable.

And there is a **CEO, CFO, or board member** whose explicit job is to ask the hard questions, verify that these three are genuinely collaborating, and understand the real risks and opportunities rather than the sanitized summary.

Notice what this structure has that the unicorn lacks: explicit accountability for each domain, no single person trying to be great at everything, and incentives that reward collaboration instead of producing role confusion.

The hiring conversation changes accordingly, and this is where most teams still get it wrong. For the Chief AI Officer, stop leading with "deep ML expertise." Look instead for someone who understands organizational design, has operated at scale, can read a balance sheet and translate a technical decision into a financial one, and can manage a diverse team. They may have never trained a model in their life. That is completely fine. For the Technical Leader, look for someone who can ship AI systems and articulate the trade-offs, with hands-on building experience rather than a theoretical background: five to eight years of doing, not of degrees. For the Governance Lead, look for someone who understands organizational risk, process, and compliance. They need to know what it means for an approval process to actually work. They do not need to understand backprop.

Then (and this is the part everyone skips) you explicitly structure the collaboration. A standing sync. Shared OKRs. A forum where disagreements get surfaced and resolved rather than buried. Not friendship. Collaboration. The difference between the two is the difference between a structure and a group text.

The Hidden Tax: What You're Actually Paying For

Suppose you get the structure right at the top. You're still probably spending far more on AI talent than you think. And not primarily in salary. You're spending it in what your best people give up while they maintain the system you built. This cost never appears in a budget line, which is precisely why it explains so many programs that invested heavily and still feel like they're getting nothing back.

Watch how it happens. You hire a strong AI/ML engineer at a critical moment. There are models to build, systems to stand up. They dive in. Six months later the initial projects ship, the models work, and the mood is triumphant. And then the job transforms. Now they're monitoring the models. Updating them. Responding when something breaks. Maintaining data provenance. Standing up behavioral monitoring. Documenting decisions. Mapping vendor dependencies. Patching governance every time something snaps.

That is not one job. That is roughly five jobs, and you have one person trying to do all of them. Your best AI talent did not sign up to be the everything person. They want to build. But they're not building: they're reading dashboards, writing incident retrospectives, and answering "why did the model do that?" for the business teams. If they had capacity to build, they would. They don't, because the infrastructure around the model demands constant care that nobody staffed for.

Here is where the math goes wrong for most organizations. They think: we need data scientists to build models, and one engineer to maintain them. What they miss is that an AI system that actually works (explainable, monitored, governed, accountable) needs far more than one maintainer. In a very small shop, those responsibilities might collapse onto one person. In any organization running more than a handful of models, it's realistically three or more. And you've filed all of it under "one ML engineer" in your budget. The reason your AI projects feel understaffed is not that AI is hard. It's that no one costed what "AI working properly" actually requires.

Then comes the burnout pattern, where the invisible cost turns into a visible resignation. Your engineer is doing five jobs and is good at maybe three. The other two are just work that has to get done, and because they're smart they can do it, but it's not what they're for, and they know it. They also know they could be doing something more interesting somewhere that would let them specialize. So they look, and they find a company that has built proper infrastructure so nobody has to be everything. You lose the person, hire a replacement, spend three months ramping them up, and they land straight into context-switching mode because nobody invested in infrastructure while the first person was here. The real cost of your AI talent was never the salary. It was the turnover, and the tax of rebuilding all that context every time someone walks.

Break down where the time actually goes and the fix becomes obvious. About 40 percent goes to **understanding**: what the models do, what they trained on, what constraints they operate under. Roughly 30 percent goes to **maintenance**: keeping systems running, responding to failures, updating models when something breaks. About 20 percent goes to **context switching**: explaining the model to other teams, answering questions, being the expert on call. And about 10 percent goes to **building new things**, which is the part everyone imagines their AI talent spends most of their time on, and the part they do least. Every one of the first three shrinks when you invest in infrastructure: explainability and provenance for understanding, monitoring for maintenance, documentation and integrated governance so context switching stops routing every model decision through the one person who understands it. Underinvest, and each becomes a constant, personal tax paid by your scarcest people.

The companies that don't have this problem all made investments that looked, at the time, excessive for the number of models they had. They versioned their data and tracked lineage, which looked like a lot of infrastructure for something that could plausibly be done by hand, right up until it meant new people could understand what trained a model without an archaeology dig. They built behavioral dashboards and automated alerting, which looked like overhead, right up until failures got caught automatically instead of by calling the one person who understands the model. They maintained decision memoranda (why was this deployed, under what constraints, what would change the decision), which looked like busywork, right up until a diagnosis that used to take days took an hour. They wrote down escalation paths and decision authority, which looked like bureaucracy, right up until governance stopped depending on a single person's memory. They paid a premium for infrastructure, and it bought them the one thing the unicorn strategy never can: their talent got to specialize, focus, and stay.

The Long-Term Math

If you're the CFO in the room, here is the arithmetic, because it favors the boring answer.

Take one AI person doing everything: building, provenance, monitoring, governance, incident response. Salary of $200K. Now go back to how that week actually divides. Roughly forty percent goes to understanding, thirty to maintenance, twenty to context switching, and about ten percent to building new things. You are paying $200K a year for a modeling capability and receiving about $20K of it. The other $180K buys work that infrastructure does better, cheaper, and without a résumé.

Now spend the money differently. Put $400K into provenance, monitoring, and governance tooling. Understanding is documented rather than reconstructed. Maintenance is alerted rather than discovered. Context stops routing through the one person who knows. Building goes from ten percent of the week to something closer to sixty. On the same $200K salary, the modeling capability you are actually buying goes from about $20K to about $120K a year.

That is $100K a year of recovered capability from one seat, against a $400K investment: on a single hire it pays for itself in four years, which is the pessimistic case, because the infrastructure is shared. Put three people on the same provenance, monitoring, and governance tooling and it returns $300K a year, and it has paid for itself before the end of year two. None of which has yet counted the second effect, which is that the job becomes interesting enough that the person stays. Replacing a senior AI hire costs a recruiting fee, three to six months of degraded output, and every piece of undocumented context that left with them. That cost is real and it is large, and you do not need it to win this argument.

The upfront infrastructure investment, the one that looked excessive and hard to justify to the board, is the cheaper option. It was always the cheaper option. It just didn't fit on a single quarter's budget line.

So if your AI engineers look overstretched, the instinct to hire more AI engineers is wrong. The answer is to invest in the infrastructure that lets the people you already have do the work you actually hired them for: provenance so understanding is documented, monitoring so maintenance is predictable, decision documentation so context stops living in one person's head, governance so the same decisions don't get re-litigated every quarter. It costs money. It costs less than the turnover of good people who burned out doing five jobs.

You Can't Hire Governance. You Build It.

There's one more altitude where the unicorn instinct reappears, and it's the most expensive place to get it wrong: governance itself. The reasoning feels airtight. You need a head of AI governance. You need a team. You need people with the right skills. So you hire. This is the wrong approach, and the enterprises that are actually good at AI governance prove it: they don't hire their way to excellence, they build capability.

The distinction matters. Hiring assumes the market contains people who already know how to govern AI at scale. Mostly, it doesn't. People know how to do one or two things: maybe fairness auditing, maybe model monitoring. But building integrated governance across fifty systems inside *your* enterprise, with your risks, your regulators, your legacy? That's not a hiring problem. That's a building problem, and no résumé solves it.

The governance capability you actually need has four components, and not one of them ships in a new hire's onboarding packet.

Governance thinking is the ability to look at an AI system and ask the right questions. Will this cause harm? Are we accountable for it? Can we defend this decision if we have to? What risk haven't we considered? This is not a skill you buy. It develops through experience: people who've governed multiple systems, learned the patterns, and can look at something new and see what's likely to go wrong. You develop it by having people govern systems over years, not by importing "senior governance people" from elsewhere. Those people know governance at *their* company. They don't automatically know yours.

Governance infrastructure is the tooling, processes, and standards that make governance scalable: monitoring platforms, decision-support systems, audit trails, governance databases. Some you buy; much you build. It requires people who understand what infrastructure you need and can design systems that work, which is not necessarily a software-engineering skill. And it comes through iteration: you implement monitoring, it fails, you redesign it; you implement decision gates, they're too slow, you automate them. Over time you have infrastructure that works because you earned it.

Governance integration is the discipline of embedding governance into how teams already build and operate, rather than bolting on a separate step. Model developers think about fairness because it's built into development. Operations teams watch for drift because monitoring is built into operations. This is not a governance-team skill; it lives in product, engineering, and operations. You build it by teaching people how governance applies to the work they already do.

Governance judgment is the ability to make trade-off decisions under uncertainty. A fairness metric slipped slightly, but retraining costs money and time — worth it? A new regulation might apply, but it's ambiguous — prepare now? A model is failing a small user segment — pull it? There's no textbook answer. Judgment comes from making hard calls and living with the consequences, over and over.

Now look at why hiring a governance team fails, because each failure mode is just one of those four components refusing to arrive by hire. The new team doesn't understand your existing systems, so they implement a generic framework that's often wrong for you. They're not integrated into how you build, so governance gets siloed — developers do their thing, the governance team checks boxes separately, and none of it changes a decision. They leave, and when they do you've lost the person but not built the capability, so you rebuild from scratch with the next hire. And they can't scale the skill across the organization, so the rest of the company still doesn't understand governance and the team becomes a bottleneck that has to review everything. A hired team is a single point of failure wearing a bigger badge.

The Multi-Year Build

Building capability instead is slower, messier, and more durable — which is a fair description of most things worth having. It doesn't live in a "governance team." It's distributed: governance *thinking* lives in your model, operations, and product teams; governance *infrastructure* lives in a platform or engineering team; governance *integration* lives inside the processes where decisions already get made; governance *judgment* lives in experienced people scattered across the organization. You might have someone coordinating it (a CTO, a governance lead), but they coordinate a distributed capability rather than running a separate function.

The build runs on the order of years, and it helps to know the shape in advance so you don't mistake slow for broken. In the first year, you develop governance thinking in your core teams: pick one system, work through governance with the team that owns it, document what you learned, then do it again with the next system. By year's end, teams are asking the right questions on their own. In the second year, you build one governance capability (monitoring, decision gates, audit trails), improve it as you learn, build the next, and start integrating governance into your existing reviews rather than adding new ones. Don't create a separate fairness review; put the fairness questions inside the model review. Don't hold separate governance postmortems; put the governance questions inside incident response. Don't add a standalone approval gate; make governance part of the deployment process you already run. By the third year your teams are thinking about governance, your infrastructure works, and you're making trade-off decisions with some maturity, so you can scale to more systems. From the fourth year on, you're scaling across the whole portfolio, and (this is the tell that it worked) you're no longer bottlenecked by a small governance team. You're bottlenecked only by how fast you can build new systems and learn from them, which is exactly the constraint you want.

Organizations that build this scale. They never hit the wall where governance becomes so much overhead that teams stop doing it and nobody flags it, and they don't lose continuity when one key person leaves. Organizations that hire a team hit that wall reliably: the team becomes the bottleneck, the team eventually leaves, and governance collapses back to zero. When you build capability, you're not building a governance function — you're building an organization that treats governance as part of how it operates. That's hard, it takes years, and it's messier than hiring a team and hoping they know what to do. It's also the only version that survives the person who built it walking out the door. One approach scales. One doesn't.

The Bottom Line

The same mistake repeats at every altitude of an AI program: you go looking for a person to embody a capability, when what you need is a structure, an infrastructure, or an organization that holds it. At the top, the unicorn hire fails because no one is world-class at model-building and organizational design and executive communication at once — you need three roles in explicit collaboration, not one adequate compromise. In the middle, a single engineer doing five jobs will burn out and leave, and the turnover costs you far more than the infrastructure that would have let them specialize and stay. And in governance, you cannot hire integrated judgment across fifty of your systems — you build it, team by team, over years. Excellence in enterprise AI is not acquired in a search. It is built, and the building is distributed and slow. That's precisely why so few organizations do it, and precisely why the ones that do pull away.

Field Kit

1. The anti-unicorn job-posting audit. Before approving any senior AI req, count how many world-class capabilities it demands in one body. If the posting requires deep ML *and* governance design *and* board-level communication, split it. Rewrite it into the three real functions: Model Builder (ships systems, hands-on, 5–8 years doing), Program Governance Architect (organizational design, regulated-industry scars), and Executive Translator (business impact, board dynamics). Approve reqs for the functions you're actually missing, not for the fantasy.

2. The role-clarity structure check. Name a specific human accountable for each of four seats: AI Program Lead / CAIO (P&L, roadmap, integration), Technical Leader (model builds, infrastructure, performance), Governance & Risk Lead (structures and accountability, may sit with the CRO/GC), and an executive sponsor (CEO/CFO/board member who asks the hard questions). If any two seats hold the same name, you have a single point of failure and a role-confusion risk. Then add the connective tissue: a standing sync, shared OKRs, and a forum where disagreements get resolved rather than buried.

3. The five-jobs diagnostic. For each AI engineer, list what they actually do this month against the five operating jobs: data provenance/lineage, behavioral monitoring, governance tracking (decisions/escalations/updates), vendor dependency management, and incident response — plus building. If one person owns three or more, you have a burnout-and-turnover risk, not a staffing win. The fix is infrastructure, not another engineer.

4. The time-allocation reality check. Ask each AI hire to estimate their split across understanding, maintenance, context switching, and building. If building is near 10 percent and the rest is consumed by the other three, you are paying senior salary for undifferentiated toil. Every point you want to move back into "building" is bought with infrastructure — explainability, monitoring, documentation, integrated governance — not with exhortation.

5. The build-vs-buy governance test. Before hiring a governance team, ask which of the four capability components you're actually trying to acquire: governance thinking, infrastructure, integration, or judgment. Thinking, integration, and judgment can only be built in your teams over years; only infrastructure is partly buyable. If your plan is to hire all four in one team, you're setting up the bottleneck-then-collapse pattern. Redirect the budget into a multi-year build: Year 1 thinking, Year 2 infrastructure and process integration, Year 3 judgment and scale.

6. The four-year TCO one-pager. For any “we just need to hire an AI person” proposal, model both paths on a single page. Path A: one person at loaded salary, turning over every 18 months, plus onboarding and lost-productivity cost per replacement. Path B: one person retained four years, plus a front-loaded infrastructure investment amortized across those years. Put both totals side by side. In nearly every case the infrastructure path is cheaper — and it’s the number that wins the room when the CFO is deciding whether “boring” is worth funding.

CHAPTER 17

Adoption: The Real Predictor of Success

Part VI · The Economics of Enterprise AI

A financial services company rolled out an AI tool that scanned customer transactions and flagged potential risks. They launched it, measured adoption, and reported success: 85% of analysts opened the tool in their first week.

Then someone asked a harder question: not whether analysts opened the tool, but whether the flags changed what they did. Everyone in the department was aware of it. Eighty-five percent opened it in the first week. Fifty-eight percent used it on a real case rather than a test one. Thirty-four percent acted on a flag at least once, which is the first point at which the tool has done anything. And twelve percent acted on it consistently, changing what they did on the system's recommendation rather than overriding it.

So: 85% used the tool once. 34% used it meaningfully. 12% actually changed their behavior because of it.

Which is the real adoption number?

Most enterprises report the first one and call it a win, and resources move on. But the only adoption that matters — the kind that changes an outcome — was 12%.

That gap, between "used once" and "actually adopted," is where most AI initiatives quietly hide their true performance.

Here is the reframe this chapter is built on.

The myth: if most of your target users are using the AI tool, it has been adopted and the investment is paying off.

The reality: usage is a vanity metric; the adoption that predicts business outcomes is whether the tool changes what people actually decide and do.

Usage is a vanity metric. The adoption that predicts business outcomes is not whether people use the tool. It's whether the tool changes what people decide and do.

Everything else is just activity.

The Metric That Lies to You

There is a number organizations watch obsessively. It's called adoption, or engagement, or sometimes just "users," and the dashboard shows it with a big upward trend line. "Our AI platform adoption is at 35%, up from 12% last quarter." And then the business outcomes don't improve. Revenue stays flat, the process doesn't get faster, decision quality doesn't shift. People are using the tool, and nothing actually changes. That is what it looks like when you measure the wrong thing.

I watched an organization deploy AI to help its sales team qualify leads. Solid accuracy, clean integration. They measured adoption by how many reps logged in each week, and the number climbed fast. "67% of our sales team is using the system." Everyone celebrated, and the lead closure rate didn't move. When they dug in, they found the reps were logging in, looking at the AI's assessment of a lead, and then doing exactly what they would have done anyway. The AI wasn't changing a single decision. It was just there. Usage was high. Impact was zero.

Here is what that teaches, stated plainly: the metric that matters isn't whether people use the tool. It's whether people make *different decisions* because of the tool. That's harder to measure (it requires connecting the recommendation to what actually happened, and understanding what the person would have done without it), but it is the only metric that tells you whether adoption is real. A recruiter can run every candidate through an AI screener and still advance exactly the candidates they'd have advanced without it: high usage, zero change, illusory adoption. Worse, the same tool can be a triumph or pure overhead at identical usage. A team that overrides an escalation-routing model 40% of the time might be exercising healthy judgment on top of a system that genuinely shifts the other 60% of decisions, or it might be ignoring a system that changes nothing. You cannot tell which from a usage dashboard.

So here is the framework that separates real adoption from usage, and it fits in two questions. First: *what decision does this AI affect?* A specific one: hire this candidate, approve this loan, escalate this ticket. Second: *compared against the decisions made without it, are the decisions different?* That requires a baseline: a holdout group, historical data, or a decision-maker you can ask what they'd have done. If they differ, in the direction the AI was designed to improve, adoption is real. If they're the same, it's an illusion, whatever the login count says.

The uncomfortable part is that real adoption plateaus faster than usage adoption. You might get 70% of the organization using the tool and only 30% of them actually changing decisions because of it. That's frustrating to report (it looks like you went backward), but it's the truth, and the truth is more useful than the appearance. If you measure usage, you celebrate when the plateau arrives. If you measure decision change, you keep investigating why it exists: maybe the model isn't good enough, maybe people don't trust how the recommendation is presented, maybe the incentives point the other way. You only find the blocker if you're measuring the thing that matters. I've watched organizations spend millions and discover, two years in, that the system never changed a single decision — while boasting excellent adoption metrics the whole time. That is a very expensive way to learn that adoption and impact are not the same thing.

Why the Gap Exists: It's the Context, Not the Tool

Once you accept that the real target is decision change, the next question is why the change so often fails to happen. The instinct is to blame the tool (it wasn't good enough, the training wasn't good enough, the business case wasn't compelling enough), so the organization doubles down on training, piles on features, or builds something else. None of those responses is exactly wrong. All of them treat adoption as a tool-or-training problem. It isn't.

Adoption in an enterprise is nothing like adoption in a consumer market, where a better product wins because people compare options and vote with their usage. Inside an organization, people don't choose based on whether a tool is best. They choose based on whether it fits how they already work, whether it creates friction with the rest of their job, whether their manager expects it, whether their colleagues use it. The best tool in the world fails if it doesn't fit the context of how the work happens.

Most adoption literature promises an S-curve: early adopters, early majority, acceleration. That is not what happens with enterprise tools. What happens is a friction curve. The first couple of months are a honeymoon: the team that built it is excited, the sponsor is excited, 50–60% of eligible people try it. Then usage flattens as people hit edge cases, and the feedback arrives: "It doesn't integrate with my process." "It's faster to do it the old way and paste the result in." "The output isn't in the format my downstream system needs." None of these is *the tool doesn't work*. Every one is *the tool doesn't work in my context*. Then leadership decides usage is too low and mandates the tool, and that is the moment you learn whether you solved the context problem or just built nice software. If you fixed the context, usage stabilizes higher. If you didn't, it collapses the instant the pressure lifts. People find workarounds, and a year later the tool is still there but no longer the default, untaught to new hires because "it didn't really take off." The standard postmortem says the tool wasn't good enough. The honest one says nobody understood the context well enough.

Context is not a polite word for "people resist change." It means the specific, structural constraints on how the work actually happens. Picture an AI model that genuinely improves sales forecasting. The reps who use it forecast better. But they're evaluated monthly on forecast accuracy, compensated on hitting the forecast, and the model updates twice a week while they submit once a week, and changing a submitted forecast requires manager approval, and the manager is traveling. The rep faces an explicit cost to using your better model against a benefit that is diffuse and shows up nowhere in their evaluation. So they don't. That is not a training problem and not a resistance problem; it's a structural misalignment between when the tool is useful and when the person is incentivized to use it. You don't fix it by training harder. You fix it by redesigning around the constraint: run the model on the person's schedule, make forecasts re-submittable without approval, or change what the evaluation rewards. None of that is about the model.

Three context failures show up again and again, worth naming because you can look for them before you build. **Workflow mismatch**: the tool works beautifully if people follow a certain path, but that path isn't how they work. Your data-prep tool assumes people upload, review, and preprocess; their real workflow is data in a database, analysis in a spreadsheet, results in a slide. So your tool sits in the middle, connects to neither end, and becomes a third tool where there were two. **Incentive misalignment**: the tool benefits the organization but not the individual. Your optimizer improves staffing across departments, but a manager's bonus rides on local efficiency, and the optimizer might move one of their people elsewhere and make them look worse. **Authority or credibility gap**: the tool recommends, but the person trusts their own judgment more. Your model recommends treatment options, accurately — but doctors won't privilege a model over their own experience until it's proven more accurate than they are, which takes time and data they don't have.

Each pattern looks different, but the structure is identical: the tool is good, and using it costs the person something (time, authority, credibility, incentive alignment) that makes not using it the rational choice. The organizations that get adoption right don't start by building the model; they start by understanding the context. They map how the work is actually done, not by asking people, who don't notice half of what they do, but by watching them work for a few days. They find the friction points, check whether the intended user is even evaluated on the thing the tool improves, and design for the context they found, building the fixes that get past the flat phase *before* they deploy, not after usage has stalled.

The Paradox: Your Best Ideas Get the Worst Adoption

Here is the part that catches technical organizations off guard: the better your idea, the more adoption risk you carry. Good ideas require people to change how they work; great ideas require them to change a lot. If your system cuts the manual work out of a process by 80%, that is a great idea and a great change-management challenge in the same breath, because a big part of someone's job just evaporated. Four hours a day of work is now forty minutes. How are they supposed to feel about that? Ambivalent, usually. Some are thrilled; some worry whether they're still needed; some liked the routine and are now bored; some were using that work to procrastinate on the hard work and now have to do the hard work. You built a great solution to a technical problem, not to an organizational one — and the organizational one determines whether anyone adopts what you made.

Struggling organizations solve this wrong. They assume the problem is education, or visibility, or trust, and build training, dashboards, and more model explanation. None of that touches the barrier. The adoption problem is solved by implementation. And implementation is not deployment. Deployment is "we built it and turned it on." Implementation is "we worked with the teams whose jobs are changing, figured out what they need to do differently, helped them do it, watched whether they did, and adjusted." It is work, and it is not automated.

In practice it means identifying whose work is changing, not "this will change the work of claims adjusters" in the abstract, but *which* adjusters, what their process is, what their incentives are. It means understanding what they *lose*, the step organizations skip: they obsess over the gains (less manual work, more accuracy) and ignore what people give up, which is autonomy, the feeling of doing skilled work, informal decision-making power. The real adoption barriers live there. It means designing the new process *with* the people who'll run it, not in a conference room: if the adjuster moves from reviewing claims to validating an AI's decisions, how much do they check, what happens when they disagree, where do they escalate? And it means rolling out in phases: 10% of claims, let the adjusters figure out what works, 25% after a month, wider after three. That's slow. It's also how you avoid the six-month "adoption is 20% of forecast" surprise.

The reason this doesn't happen more often is uncomfortable. It requires change-management people, not change-management theory: people who can spend a week sitting with a team of adjusters, understand deeply what actually worries them, and build the solution together. That is expensive and demands interpersonal skill some technical teams find genuinely uncomfortable, so organizations underinvest. They hire a consultant who produces a fifty-page plan, and a fifty-page plan is not implementation. It's theater. Implementation is the unglamorous, in-person, week-after-week work of helping specific people change how they work. Most AI projects don't fail because the technology is bad. They fail because implementation was an afterthought. You built the machine, the machine works, and you forgot to teach people how to work with it. No amount of technical excellence fixes that.

The Adoption Threshold: Measuring It Honestly

If usage is the wrong metric, you need a definition of adoption sharp enough to act on. The common enterprise definition is binary (did you use the tool, yes or no), which produces meaninglessly high numbers that mask real problems. Actual adoption, the kind that changes behavior and produces value, is a ladder with five rungs:

1. **Awareness**: users know the tool exists and what it does.
2. **Accessibility**: they can actually use it in their workflow; it isn't one more login, blocked by process, or harder than the alternative.
3. **Trust**: they believe the output is correct and relevant to their decision.
4. **Integration**: using the tool is genuinely easier than not using it; it saves time or produces a better decision.
5. **Behavior change**: they actually change what they do based on the recommendation.

Most tools clear the first three rungs and stall on the last two. Rung four is where adoption usually dies: the user opens the tool, reads the output, and thinks "this would help, but it's faster to just decide myself." The tool is present in the workflow and absent from the decision. Rung five is where you find out whether any of it mattered. Plenty of people will use a tool and override it every time, "using" it without ever letting it change what they do. A fraud analyst who overrides the system 60% of the time is using it but not trusting it; a salesperson who reads a low lead score and calls the lead anyway, because the old rule was call everyone who inquired, is using it but not acting on it; a moderator who falls back on personal judgment at every edge case is using it but not relying on it. In every one of those cases the system is a second opinion rather than a decision — and second opinions don't show up in outcomes.

So measure the things that distinguish those cases. Start with frequency, not first use: weekly or more is meaningful, monthly is marginal, one-time is marketing. Then the action rate: when users see the output, what fraction of the time do they act on it rather than override it? Consistent overriding means the system isn't trustworthy, or it isn't solving the problem users actually have. Then behavior change directly, by comparing genuine adopters against non-adopters; if adopters don't outperform, adoption isn't producing the value it should. Then run the removal test: if you switched the tool off on a Monday, would anyone's Tuesday look different? If the honest answer is no, you have supplementation rather than reliance (an extra input people consult and set aside), and no amount of usage growth converts that into value. Finally, when people open a tool and abandon it, ask why, because a trust problem, an integration problem, a relevance problem, and a process problem each demand a different fix, and you can't tell them apart from a usage chart.

That gives you a practical threshold for calling adoption real: at least 50% of target users using the system weekly, at least 60% of those acting on its output more than half the time, and the adopters measurably outperforming non-adopters on the outcome that matters. Clear that bar and you have adoption that changes behavior at scale. Fall short and you have a tool some people open that doesn't change what anyone does. That's worth diagnosing before you spend another dollar on a better model when the real problem was never the model. The reframe is not that tools must be perfect before deployment; it's that you measure adoption correctly and stay honest about what the numbers mean. "85% of users opened the tool" is a marketing success. "34% of users are using it meaningfully" is the true conversation, and only the second one tells you whether to invest in the model, the workflow, or the trust gap. Most enterprises skip the measurement step, report the flattering number, and declare victory. The tool never changes behavior at scale.

Launch Is Not Adoption

There's one more place the usage illusion hides, and it's in time. A model goes into production on a Tuesday. There's a deployment meeting, a celebration, a Slack message: "Model X is live, nice work everyone." No errors, predictions flowing, infrastructure stable. Everybody treats that as the finish line — and the adoption question, the only one that determines whether any of it mattered, hasn't been asked yet.

Software has a clean success moment. You merge, you deploy, nothing throws an error, you won. AI doesn't work that way. You already know the quiet version: Chapter 12's model that keeps every light green while its decisions walk somewhere nobody intended, surfacing months later through a customer or a regulator rather than a dashboard. That's the failure where the system degrades. This chapter is about the other one, at least as common and far less discussed: the model that stays perfectly accurate for two years while nobody's work changes because of it.

So the people who know what they're doing don't celebrate at launch. Launch is when exposure begins, not when success is determined. A deployment actually succeeds three to six months later, when six conditions hold — and the two that decide whether the money was well spent have nothing to do with your monitoring stack.

The first is **ownership**. At launch, the team that built the model is watching it, and they know what normal looks like because they defined normal. That isn't a running system; it's a supervised experiment with a good dashboard. Success means ownership has actually moved: ops or the product team holds the alerts, the build team is on call rather than on duty, and the runbook was written by someone who had to ask questions to write it. The test is blunt: if the person who wrote the training pipeline took three weeks off, would anybody notice the model was unattended? If yes, you don't have a deployed model. You have a loan.

The second, and the most overlooked, is that **the people it was built for have changed what they do**. A model can be technically correct, well monitored, cleanly owned, and still a failed deployment if the product managers route around it because they don't trust it, or the analysts read the output and then do what they were going to do anyway. This is the same decision-change test the rest of this chapter has been making, applied later in the calendar. And the timing is the point. Behavior change in week one is enthusiasm; the tool is new and the sponsor is watching. Behavior change in month six, after the novelty is gone and the sponsor has moved on to the next initiative, is adoption.

The remaining four are operational, and Chapters 12 and 13 cover them properly: monitoring that catches the failures that matter rather than just the errors; a real process for acting on an alert, because monitoring nobody acts on is decoration; a season of input change survived, because a model that has only ever seen January doesn't know what it does with February; and at least one real problem found and fixed in production. They belong in an adoption conversation for one reason: each is a way the tool stops deserving the trust you spent months building, and trust is asymmetric. Users who get burned once by a stale recommendation don't come back at the rate they left.

This is what most governance gets wrong, because most governance is built around launch: review the model, approve it, ship it, done. A mature process doesn't end at launch. It shifts. Approval gates give way to monitoring gates, ownership passes to ops, governance settles into periodic review. So when someone brings you a deployed model and says "we're done," treat it as the beginning of the real work, and ask the questions that have checkable answers: has ops actually taken ownership, or is the build team still babysitting it? Can we run this without the person who built it? Have we seen the inputs shift and confirmed the model still holds? Have we acted on a live alert and fixed a real problem? And then the one this chapter exists for: has anybody's behavior changed? Every answer yes, the deployment succeeded. Any "not yet," you're still deploying — you just haven't finished.

What Actually Works

Pull the threads together and one discipline emerges, and it isn't a technical one. The organizations winning with AI are not the ones building the most sophisticated models. They're the ones who treat adoption as something to be designed rather than hoped for, and who put a named person on it before anyone writes code.

In practice this looks distinctly unimpressive. It looks like a program lead I know who won't let a tool past 25% rollout until she can put two numbers on the same slide (what adopters do and what non-adopters do, on the outcome the business cares about) and who, when those numbers match, sends someone to find out why instead of sending an email about usage. It looks like tracking the trust trajectory rather than the usage trajectory: are people relying on this more than they did in month one, or rebuilding the old workaround? A decline there is a defect report, not a training problem.

The reason to do this work is that the alternative compounds against you. An abandoned tool doesn't just waste its development cost, its infrastructure spend, and the training hours; it spends the organization's willingness to try the next one. The team loses credibility, the sponsor turns hesitant, and eighteen months later your genuinely good proposal walks into a colder room than it deserves. You can survive one AI tool nobody uses. What's expensive to rebuild is the assumption that when your team says something will change how the work gets done, it will.

FIGURE 17.1

The Adoption Threshold Funnel

Five thresholds between the launch email and a decision that came out differently.

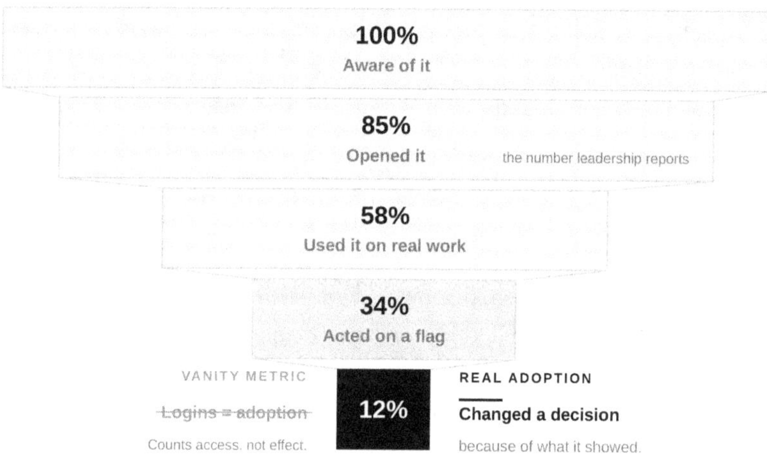

Stage five is the only stage that pays for anything.

Leadership quotes stage two and calls it adoption. Six of every seven people who opened it never changed a single decision because of it.

Workshop use: each team writes its own five numbers into the bands, in order. The login metric is retired on the spot; stage five becomes the reported figure.

FIGURE 17.1

The Bottom Line

Usage is a vanity metric. The adoption that predicts business outcomes is behavioral: does the tool change what people actually decide and do? Opening a tool is not using it, and launching a model is not succeeding with it: at every stage the flattering number counts activity while the useful one asks whether a decision came out differently because the system existed. Most enterprises measure logins and first-opens, report the impressive figure, and never notice that the tool changed no decisions, cleared no friction, and survived no season of real input. Real adoption is a context problem, an implementation problem, and a measurement problem long before it's a technology problem — and the work of solving it is slow, in-person, and boring. Measure decision change instead of usage, design for the workflow people actually have, and treat launch as the moment the real work begins, not the moment it ends.

Field Kit

1. The two-question adoption test. For any AI tool: (a) *What specific decision does this affect?* (hire/loan/escalate/route). (b) *Compared against decisions made without it, are the decisions different — and different in the direction we intended?* If you can't answer (b), you don't yet know whether the tool is adopted. Build a baseline (holdout group, historical comparison, or ask the decision-maker what they'd have done) before claiming success.

2. The real-adoption threshold (report these three numbers, not logins). - ≥ 50% of target users use the tool at least weekly. - ≥ 60% of those act on its output more than half the time (rather than overriding it). - Adopting users measurably outperform non-adopting users on the outcome that matters. Clear all three and you have adoption at scale. Miss any, and diagnose before you scale. Sanity-check with the removal test: if the tool went dark on Monday, would anyone's Tuesday look different?

3. Pre-build context map. Before writing code: (a) map how the work is done today by *watching*, not asking; (b) mark every friction point where the tool adds a step or needs credibility the user hasn't earned; (c) check whether the intended user is even evaluated on the thing the tool improves; (d) name what the user *loses* (autonomy, status, informal power), not just what they gain.

4. Non-adoption interview script. When someone opens the tool and stops, ask which of four it is, because each has a different fix: Trust ("I don't believe the output") → better model / explainability. Integration ("it's too slow this way") → redesign the workflow. Relevance ("it doesn't answer my real question") → you misunderstood the need. Process ("my old way is faster") → make the tool the path of least resistance.

5. Phased rollout default. Don't turn the model on for everything. Start at ~10% of volume, let users figure out what works, go to ~25% after a month, widen after three — and pre-load the fixes that get past the friction phase *before* deployment, not after usage stalls.

6. "Are we actually done?" deployment gate. A model is only successfully deployed when you can answer yes to all five: Has ops taken ownership (or is the build team still babysitting it)? Can we run it without the person who built it? Have we seen the inputs shift and confirmed the model still holds? Have we acted on a real monitoring alert and fixed a real problem? And the one teams skip: six months in, with the sponsor's attention gone, has anybody's behavior actually changed? Any "not yet" means you're still deploying — keep governing.

CHAPTER 18

Measuring What Matters: Metrics, Dashboards & ROI

Part VI · The Economics of Enterprise AI

The recommendation engine still runs. It still hits its precision target. Nobody has acted on its output in five months.

Every data science team has a graveyard, and it is not on the org chart or in the budget forecast. It's the projects that worked on paper. They hit their targets. The models converged. The metrics looked clean. Then, four months into production, without anyone quite noticing when, they became optional.

A recommendation engine no one acts on anymore. A churn model with good precision that the business stopped using because it "doesn't feel right." A customer segmentation model that's accurate by every measure except the one that matters: whether anyone can do anything useful with the segments.

Ask the team what went wrong and they'll tell you the truth: nothing. The metrics didn't move. The model didn't degrade. Everything performed as expected.

That's the problem.

The wrong metrics don't fail. They persist. They hum along, proving they work, while the business stops trusting them. A model that fails loudly gets fixed. A model that succeeds on the wrong metric gets funded for another year.

This chapter is about the gap that fills that graveyard: the gap between the numbers your team measures and the numbers that decide whether your AI program lives or dies. Model metrics and business value are different languages. Most enterprises are fluent in the first and illiterate in the second, and they don't find out until the CFO stops approving the follow-on investment. The work of closing that gap is not glamorous. It isn't a better model. It's an information architecture that connects what you measure to the decisions someone with a budget actually has to make.

Here is the reframe this chapter is built on.

The myth: a model that hits its accuracy targets is working, and if it ever stops delivering value the metrics will tell you.

The reality: the wrong metric persists instead of failing, so measure the business outcome and wire it to an explicit decision with named criteria and a named owner.

You're Measuring at the Wrong Level

Most enterprise teams measure AI success through a lens built for a different problem: can we build a model that predicts X with accuracy Y? That's a data science problem. It is not an enterprise problem. An enterprise problem sounds like: can we deploy a system that helps a human make a better decision, at a cost we're willing to pay, in a way that creates value we can point to? These are not the same question. You can answer the first one brilliantly and lose the second one without a single metric ever turning red.

Here's how it plays out. Your team builds a credit risk model. They measure it on hold-out data. Precision, recall, AUC: all beautiful. 0.92 AUC. You can draw the ROC curve and it's smooth and clean. Then it goes live, and in month two a credit analyst pulls you aside: "The model is right more often than I am. But I'm the one who has to justify the reject to the customer. When the model says no and I say no, I sound like a robot reading numbers. I can explain it in a way they trust."

Suddenly the 0.92 doesn't matter. What matters is whether the analyst uses the model or works around it. Most organizations don't measure that. It isn't a metric — it's an "adoption problem," and adoption problems get blamed on the humans. "If they'd just trust the model more." But the humans understand the metrics perfectly. They're answering a different question: does this reduce my risk while I do my job?

The real cost shows up later. You've deployed 47 models. By your metrics, 44 of them work. By the organization's metrics (are they actually used in production decisions, are they creating value, do people trust them), maybe 12 work. You've spent millions on infrastructure, data science teams, and governance frameworks, and built a technically sophisticated machine that delivers technically perfect predictions nobody cares about. This is why so many AI programs feel expensive and slow and never quite justify the investment. It isn't incompetence. It's measuring success at the wrong level.

The Metrics Your CFO Actually Cares About

You have metrics. Good ones, probably. Your recommendation model has a 94% NDCG. Your classifier has an F1 of 0.87. You track precision, recall, AUC. You measure fairness. You have dashboards.

Your CFO does not care about any of it.

Not because your metrics are wrong — because your CFO cares about different metrics, and most technical organizations miss the gap. When you say "our model improved precision by three points," what a CFO hears is "we spent six months optimizing something abstract." Ask how much money that's worth and you don't have a good answer. That disconnect, between model performance and business performance, is where most projects get trapped.

Your CFO cares about four things, and none of them is model F1. **Conversion:** did the initiative produce more revenue or cut cost? **Risk avoidance:** did it prevent something bad from happening, and how much? **Velocity:** did it let you do something faster, so you could do more volume? **Margin:** did it improve the margin on business you already have?

Each is measurable. Consider a recommendation model. Your metric is "NDCG improved from 0.72 to 0.75." Your CFO's metric is "users who saw the new recommendations bought 6% more than users who saw the old ones." Different thing entirely. The NDCG lift was necessary to get there, but it isn't what the business is buying. Or a fraud model. Your metric: "true positive rate up from 87% to 91%." Your CFO's metric: "we caught $2.3 million in fraud last quarter, up from $1.8 million the quarter before," and, on the other side of the ledger, "our false positives cost us $640,000 in the same quarter, in manual review hours and abandoned checkouts." That second number is Chapter 6's error-cost worksheet with a dollar sign and a reporting period attached. The translation work is the whole job: same errors, stated in the currency Finance already budgets in.

The pattern holds every time. Your metrics are leading indicators. The CFO's are lagging indicators. And lagging indicators are what determine whether you keep getting funded.

Why do technical and business teams end up measuring different things? Four reasons, and they compound. Your metrics are *available*: you're logging predictions and holdout accuracy; you're not usually inside the business system where you can see whether a recommendation turned into a purchase. Your metrics are *crisp*: "F1 of 0.87" is unambiguous, while "the model prevented fraud" requires attribution models, control groups, and arguments about whether it was the model, the sales team, or a pricing change that moved the number. Most teams *assume model performance maps to business performance*: usually true, not always, and the gap between a technically excellent model and one that actually moves the needle is where projects stall. And measurement is *political*: if you own the model you want metrics that show it in a good light; if you sit in the business you want metrics that show you created value; when those differ, it's safer to retreat to model metrics than to fight about attribution.

The fix isn't hard, but it has to be deliberate. Start from the business question. Not "is the model accurate," but "what does accurate actually deliver: more revenue, less risk, faster processing, happier customers?" Then translate that into an outcome you can test. "Should we use this model for loan approvals" does not resolve to "F1 > 0.85." It resolves to "loan defaults decrease by 5% and loan volume increases by 10% when we use the model versus traditional underwriting." Now you have something that matters to the CFO. The way you get there is an A/B test: put the model in production for a subset of transactions, and measure not "does the model work," but "does deploying it change behavior in the direction we want?"

That reframing changes the conversation. Walk into the CFO's office with "I want to build a demand forecast model, and here's what we'll measure: does it reduce inventory carrying cost by more than the model costs to run?" And you have a conversation. The CFO can push on what the model costs, what typical carrying cost is, whether you're confident in those numbers, how you'd validate it. You're speaking the same language. Walk in with "the model has an RMSE of 12.3%" and the CFO does not know how to respond. She doesn't know if that's good. She doesn't know if it's worth investing in. She says "sounds nice, let's see if it works," and six months later you're having a frustrated conversation about why the forecast project never got follow-up funding.

So here is a diagnostic for any project that feels stuck. Can you answer, quickly and concretely: *if we deploy this model, what business outcome will improve, and by how much?* If you can't, that's not a technical problem — that's why it's stuck. If you can, and the CFO is still skeptical, that's a different and much healthier conversation. That's a risk conversation, and at least it's about something real.

Three Layers, Built Upside Down

A retail company deployed a recommendation engine and tracked its success with these metrics: API latency, model inference time, recommendation coverage (the share of the catalog the model could recommend), and accuracy on a holdout test set. By every one of them, the system was excellent. Latency was subsecond. Inference was fast. Coverage was 98%. Accuracy was strong.

Then they checked the business metrics. Recommendations clicked: 4%. Recommendations purchased: 0.2%. Revenue impact: negligible.

The system was technically perfect and operationally irrelevant.

This is not a story about bad metrics. Every number on that dashboard was real and worth knowing. It's a story about measurement philosophy, about confusing activity metrics with outcome metrics, and then wondering why good metrics produce bad results. Technical metrics feel like they should translate into business impact. Better accuracy seems like it should mean more clicks; faster latency seems like it should mean more sales. Sometimes that's true. Often it isn't. The relationship is weaker than intuition insists.

A system that produces business value needs three layers of measurement, and the order matters more than the contents.

Layer 1 is business outcome. Is this delivering what we wanted from a business perspective? For a recommendation engine: click-through rate, conversion rate, revenue per user. For a fraud system: the dollar value of fraud prevented, and the false-positive rate that drives operational cost. For a churn system: customers retained who would have churned, and cost per retention interaction. For a pricing system: revenue, margin, elasticity impact. These are the metrics that matter. Everything else is instrumental to them.

Layer 2 is user behavior. How are people actually interacting with the system? Impressions, clicks, dwell time, the conversion funnel; review rates and override rates; adoption of recommended prices. Layer 2 is the bridge between technical performance and business outcome. If Layer 2 is healthy but Layer 1 is flat, you have learned something precise: the problem isn't model quality, it's something about the user experience or the business context around it.

Layer 3 is technical. Is the system functioning as designed? Latency, throughput, error rate; accuracy, precision, calibration; coverage, consistency, infrastructure cost. Layer 3 is a prerequisite for value. It does not determine value.

Here's the trap, and nearly every enterprise falls into it. The common pattern: deploy the system, measure Layer 3, Layer 3 looks good, check business impact six to twelve months in, business impact disappoints, blame the model, build a better model, go back to measuring Layer 3. The loop never works, because the model was rarely the problem. Either users weren't using the system at all (an adoption problem), or they were using it and it didn't solve their actual problem (a fit problem), or the predictions were correct but something else in the business (pricing, logistics, timing) blocked the impact.

The fix is to build the ladder upside down from the way instinct wants to build it. Start at Layer 1, *before* you deploy: what would it mean for this system to deliver business value, what would you measure, what's the target? For some systems this is genuinely hard: the impact is indirect or long-term. Do your best; it is far better to measure something imperfect than to measure nothing. Then measure Layer 1 continuously and treat it as your north star. Use Layer 2 as a diagnostic: when Layer 1 is flat, Layer 2 tells you whether the problem is adoption or value. Layer 3 you're probably already measuring. The discipline is to keep it a supporting metric, not the driving one.

That gives you a diagnostic ladder instead of a guessing game. Layer 1 good? Don't optimize further; you're done. Layer 1 bad, but Layer 2 shows people aren't using the system? That's a product and adoption problem (awareness, UI, timing), not a modeling project, and no amount of retraining will fix it. Layer 2 shows people using it but getting no value? *Now* you look at Layer 3, because now the technical quality might genuinely be the bottleneck: the recommendations are bad, the timing is off, something in the system is broken. The framework's entire purpose is to stop you from solving the wrong problem, from perfecting a technical metric while business impact sits flat.

The cost of setting this up is low: a few hours of honest thinking about what business value actually means and how you'd observe it. The payoff is high. You find out whether your system is delivering value, and when it isn't, you can say why instead of blindly optimizing the wrong metric.

From Measurement to Decision

Suppose you do all of this. You measure the right layers. Your dashboards are comprehensive. You still may not know what to do.

Chapter 2 settled the principle: every "we will monitor X" is unfinished until it reads "and if X exceeds threshold Y, [named person] decides whether to accept or remediate it." Take that as given. What nobody warns you about is the distance between writing that sentence into a policy document and getting it onto a screen an on-call engineer looks at on a Tuesday afternoon. Closing that distance is what this section is about, and it is where most governance programs stop.

Because the failure you graduate into, once you've solved the first one, is not too few metrics — it's too many, disconnected from decisions. You can have perfect metrics and still not know what to do about them. You can have comprehensive dashboards and still be paralyzed about which system to prioritize. The gap between measurement and action is where good governance goes to die.

Watch it happen. A model's fairness score degrades. An alert fires. Then what? Retrain? Pull the model? Investigate why the score dropped? Different teams answer differently, so the decision gets deferred. Meanwhile the data science team has its metrics (accuracy, latency, feature importance) sitting on a dashboard people glance at occasionally. The business team has its own (satisfaction, churn, revenue) and doesn't connect them to the model. The governance team has fairness scores and bias audits, monitors them faithfully, and when something goes out of bounds doesn't always know what to do. Three teams, three dashboards, no decision.

Measurement should feed four things, and if it doesn't, it's just overhead.

The first is **alerting and escalation.** Most organizations have the alert and not the authority. Someone learns that fairness degraded but can't investigate, doesn't have the resources, and has no criteria for deciding what to do. The alert fires and nothing happens.

The second is **trade-off visibility.** Your model makes trade-offs: accuracy against fairness, latency against cost, interpretability against predictive power. Most organizations measure accuracy and measure fairness but never measure the trade-off *between* them, so when they need to optimize one at the expense of the other, they do it blind, without the data to make the choice deliberately.

The third is **portfolio-level prioritization.** You have 50 systems, all slightly out of compliance, all improvable. Which do you fix first? That requires measuring impact across systems: which affects the most customers, which carries the most risk, which is degrading fastest. Most organizations have system-level dashboards and no portfolio view, so prioritization happens by politics instead of by data.

The fourth is **root-cause understanding.** When a score drops, you need to know why: did the training data change, did the input distribution shift, did the model start leaning on a new feature, did the training process break? Most organizations know something is wrong and not why.

Closing that gap is a matter of building decision architecture, which is less exotic than it sounds. There are four moves. **Define the decision points** for each system: retrain or not, pull or not, investigate the anomaly or not, optimize for accuracy or fairness, add human review or not. Make them explicit. **Define the decision criteria:** when accuracy drops below X, investigate; when the fairness gap exceeds Y, escalate; when latency spikes above Z, consider pulling. This sounds obvious, and almost no one does it, which is why decisions get made inconsistently or not at all. **Define the decision authority:** not a committee, a person or a role, with a named escalation path for when they need help. And **connect measurement to criteria** so the dashboard highlights when a criterion is met. You shouldn't have to stare at twenty metrics and infer what they mean. The dashboard should say: "This metric triggered decision criterion X. Recommendation: escalate to Y." That's the shift: building measurement and dashboarding as decision support, not as analytics.

The difference is stark in practice. A model's accuracy degrades from 92% to 87%; the alert fires. Without decision architecture, it lands in Slack, someone looks at it, has no criteria for whether 87% is acceptable, has no authority to act, mentions it in standup, and nothing happens while accuracy keeps sliding. With decision architecture, the dashboard reads: "Accuracy degraded from 92% to 87%. Decision criterion met: accuracy below 88%. Recommendation: investigate root cause within 48 hours. If the cause is data quality, escalate to the data team lead. If it's the model, escalate to the model owner. If it's unclear after 48 hours, pull the model and switch to fallback." Now the on-call engineer knows what to do, has the authority to do it, and something happens.

Governing at Scale

Decision architecture is what lets governance scale without scaling headcount, and it is the one piece that survives intact through every rebuild threshold Chapter 22 walks you through. The mechanism is boring, and it is the whole trick: decision points repeat, so the architecture compounds.

Concretely. You write a criterion for the credit model: fairness gap above four points, escalate to the model owner within five business days, pull to manual underwriting if it isn't explained inside ten. Then the collections model needs the same criterion with a different number. Then the pricing model needs it too. Fairly quickly you stop authoring one decision architecture per system and start maintaining four or five recurring *decision types* (degradation, drift, fairness gap, cost overrun, coverage collapse), each with a settled escalation path and a named role at the end of it. Onboarding a new system becomes an afternoon of filling in thresholds and naming owners. Rebuilding the reasoning from scratch for each system takes a quarter, and that is what most teams do.

Everything people call advanced falls out of that. Rules that escalate on their own are a small engineering job once the recurring decision types are written down — you're automating a decision you already made, not inventing one. A portfolio view (which system is degrading fastest, which touches the most customers, which has the thinnest fallback) is the same criteria read across rows instead of down a column. Neither one is a new measurement program, which is why neither one needs a new headcount request.

Notice what does not change. You are not adding more measurement. You already have the data you need. You're connecting the data you have to decisions more systematically. The enterprises that do this well end up with something that looks like automated governance. They don't have more people. They have better decision architecture.

So start with one system. Map its decision points. Define the criteria and the authority. Build the dashboard. Then test it against the next real problem: did you catch it early, and did the right person make the right call? Replicate for the next system, and watch for the decision points that recur: those become your common criteria and your shared dashboards.

What Actually Works

Put the four movements together and a single discipline emerges, and it is not the one enterprises reach for first. The organizations that demonstrate AI ROI are not the ones with the best data scientists, and not the ones with the best models. They're the ones that were clear about what success meant before they started (clear in a way that was measurable, observable, and tied to something the business cares about) and then had the nerve to act on the answer.

That discipline looks like a head of AI I know who runs a quarterly check on every model in production. It's three questions: Is anyone using this? Are they using it the way we expected? Is it creating value? If any answer is no, they either fix it or retire it. No heroes. No "it's technically correct, so we must be doing something else wrong." Just: does this work in practice? A team willing to retire a model that scores well but delivers nothing will beat a team that keeps polishing the AUC on a model no one uses.

None of this requires better technology. It requires deciding what success means before you build, measuring the outcome and not just the model, and wiring your measurements to the decisions someone with a budget actually has to make. Do that, and the model metrics take care of themselves.

The Bottom Line

Model metrics and business value are different languages, and the wrong metric doesn't fail loudly — it persists, humming along while the business stops trusting it. Measure in three layers, but build the ladder upside down: start from the business outcome your CFO can recognize, use user behavior to diagnose why an outcome is flat, and treat technical metrics as support rather than the score. Then close the gap most enterprises never close: wire every metric to an explicit decision, with named criteria and named authority, so an alert produces an action instead of a Slack message no one owns. This is not a modeling problem, and the answer is not better data scientists; it's the boring discipline of defining success before you build and retiring what doesn't deliver. Done well, it looks like automated governance — not more people, just better decision architecture.

Field Kit

1. The CFO conversation starter (question set). Bring these to any funding or kickoff meeting, before a line of code is written: *If we deploy this, what business outcome improves, and by how much? Which does it move — conversion, risk avoidance, velocity, or margin? What does the model cost to build and run, and how big is the thing it improves? How will we validate the number — A/B test, control group? What's our leading indicator, and what's the lagging indicator the CFO will actually judge us on?* If you can't answer the first question quickly and concretely, the project isn't stuck for technical reasons.

2. The three-layer metrics map (one-page template). For a single system, fill three rows. *Layer 1 — Business outcome:* the metric, its target, and how it's measured. *Layer 2 — User behavior:* adoption, usage, override/review rates. *Layer 3 — Technical:* latency, accuracy, calibration, cost. Hard rule: you may not deploy until Layer 1 has a named target. If Layer 1 is blank, you're about to add a system to the graveyard.

3. The quarterly model review (three-question diagnostic). For every model in production: *Is anyone using this? Are they using it the way we expected? Is it creating value?* Any "no" means fix or retire. Track the retire count — a program that never retires anything isn't measuring outcomes, it's just accumulating them.

4. The diagnostic ladder (decision aid for disappointing results). Layer 1 good → stop, you're done. Layer 1 bad *and* Layer 2 shows low usage → adoption/product problem, not a modeling project. Layer 1 bad *and* Layer 2 shows usage but no value → now, and only now, inspect Layer 3 for a technical cause. Use it to stop teams from rebuilding a model when the real bottleneck is the UI, the timing, or the fact that no one knows the system exists.

5. The decision architecture checklist (per system). (a) Decision points listed — retrain, pull, investigate, add human review. (b) Criteria set with numeric thresholds — below X → investigate; gap > Y → escalate; latency > Z → consider pulling. (c) Authority named — a role, not a committee, plus an escalation path. (d) Dashboard surfaces the *triggered criterion and recommended action*, not twenty raw metrics.

6. The alert-to-action template (one line your dashboard must be able to produce). *"[Metric] moved from A to B. Criterion met: [threshold]. Recommendation: [action] within [time window]. Escalate to [role] if [condition]."*

If your alerts can't be written in this form, they aren't connected to decisions yet — they're noise.

PART VII

Compliance, Law & the Regulatory Future

CHAPTER 19

What Regulators Actually Expect

Part VII · Compliance, Law & the Regulatory Future

A regulator walked into a bank's AI governance office and asked a simple question: "Show me your decision log for the AI systems you're using in lending."

The bank was ready. It produced a clean spreadsheet: model name, version, accuracy metrics, deployment date. Everything a modern model risk program is supposed to have.

The regulator looked at it and asked again. "Where's the decision log? When did you decide to use this model for this decision? Who approved it? What was the alternative you rejected, and why? When the output changed, who signed off on the change?"

Blank looks.

What the bank had built was a model registry. What the regulator wanted was a decision record. Those are not the same document, and the gap between them is the gap most enterprises are about to fall into.

Here is the reframe this chapter is built on.

The myth: regulators are coming after your AI with specific technical requirements.

The reality: they're coming after you for governance failures that AI just happens to amplify.

Over the past year I've talked with regulatory and compliance leaders at a dozen organizations, and the conversation is remarkably consistent. They're building AI governance programs that look sophisticated: model risk frameworks, fairness audits, documentation templates, validation suites. But when they actually sit across from a regulator, in an exam or an inquiry or a pre-filing consultation, the questions are almost never about the AI itself. They ask: How do you know what your model is doing? Who decided this was acceptable to deploy? What happens when something goes wrong? How would you even know it had degraded if you weren't looking? Did you consider this particular risk? Can you actually explain the decision it made in that specific case?

This is not a prediction about temperament. When the EU moved its own high-risk deadline in July 2026, the reason it gave was that conformity assessment bodies and harmonized standards were not in place, which is to say the regulator delayed the audit because the audit infrastructure was not ready, not because the underlying expectation had softened. The expectation is the same one it has always been. Show your work, show who decided, show what they knew.

None of those are technical AI questions. They're governance questions. They're accountability questions. They sound exactly like the questions a board asks about any high-risk business process, because that is precisely what a regulator thinks your AI is. Not a technical marvel. A high-risk business process that happens to run on math.

Most organizations have this backward. They're building technical AI governance as though the regulatory risk were technical. It isn't. The regulatory risk is organizational. Did you think about it? Did you decide it was acceptable? Can you prove it?

They're Regulating the Decision, Not the Model

Start with what regulators don't care about, because it's probably not what you've been told.

They don't care whether you used logistic regression or a transformer. They don't care about your accuracy metrics. They don't care whether you hit parity on your fairness score. They don't even care, in the strictest sense, whether your model is biased. They care whether you *knew* it might be biased, and whether you *decided* that was acceptable.

Sit with that distinction, because it's the whole chapter. Regulatory failure is not "we had a biased model." Regulatory failure is "we had a biased model and either we didn't know it could happen, or we knew it could happen and we deployed it anyway without disclosing it or managing it." The model being imperfect is not the violation. The decision being made without appropriate information and oversight is the violation.

Regulators work from an old and simple assumption: you have obligations to your customers, your shareholders, the public. You made a decision to deploy AI. If that decision creates a risk of unfair treatment, customer harm, financial loss, or a compliance breach, you are accountable for the decision. Not for the model being perfect. For the decision being deliberate.

This is why compliance teams keep feeling like AI governance is a moving target. They're being asked to regulate decision-making, not technology. You cannot write a rule that says "make good decisions about AI" the way you can write one that says "maintain a capital ratio of X." So regulators fall back on the only question available to them: how do you know you're making good decisions? Document your process. Show your thinking. Prove you considered the risk before you took it.

The framework a regulator brings into the room is decades old. It's the same one they use for credit decisions, lending practices, capital allocation, insurance underwriting, hiring. Does the organization have a clear governance structure: an actual person accountable for the deployment, not a committee that reviews audits after the fact? Did it do a risk assessment before deployment, not "was the model tested," which is validation, but "did you consider that this could produce unfair or incorrect outcomes, and decide the benefit was worth it?" Are there clear criteria for continuing to operate it: what threshold or observation would make you pull it? Is there evidence of real oversight: a human with responsibility actually paying attention, rather than an automated system someone switched on and forgot? And can you explain a specific decision: pick a customer outcome, tell us why it happened, trace it through your process?

Different regulators weight those five differently. Some lean hardest on fairness, some on security, some on operational resilience, and the emphasis shifts by jurisdiction and by the philosophy of whoever is holding the pen. All of them care about the decision. And the fact that a machine is in the loop changes none of it. The regulatory expectation is the same one that has governed high-risk decisions for a century: you thought about it, you decided it was acceptable, you wrote down your reasoning, and you're watching it.

Where the Money Goes to the Wrong Place

Once you see that the risk is organizational, you start noticing how much governance spending defends nothing.

I've watched organizations pour real resources into activities that, if a regulator actually pushed on them, would do almost nothing to defend the underlying decision. They build fairness metrics with no process for deciding whether a given level of fairness is acceptable. You measure the disparity, but who signed the line that says "this level is okay, and here's why"? Nobody did. You built a dashboard. They create model documentation that describes in exquisite detail what the model does but never why the business decided it was worth doing, so when the regulator asks "who approved deploying this knowing it would make this particular call incorrectly three percent of the time," there's no answer, because it was never framed as a decision. They stand up validation frameworks that prove the model works but not that it's solving the right problem. They run fairness audits after deployment and call it governance, when an audit is a reaction: it tells you what went wrong; it doesn't tell a regulator you thought it through before you took the risk.

None of these activities are wrong. They're supporting infrastructure. The mistake is treating the infrastructure as the building. The core governance question underneath all of it is singular: did this organization make this decision deliberately, with appropriate oversight, knowing the risks, and documenting the reasoning? If the answer is no, every fairness metric and validation suite you own is finish work on a house with no foundation.

Here's where that leaves most enterprises specifically exposed. On **decision documentation**, you can explain what the model does but not why anyone in your building decided it was acceptable to deploy a model that makes this call this way. On **ownership**, ask "who is accountable for this system?" and if the honest answer is "the data science team," you've just revealed you don't have governance: a technical team is babysitting a system, and nobody with P&L accountability owns it. On **alternatives**, you deployed the AI, but did you consider not deploying it, or using a simpler model, or keeping a human in the loop? You don't have to choose the lowest-risk option. You do have to show you knew it existed. And on **thresholds**, when exactly do you pull this thing, not theoretically, but the actual number of complaints or the specific drop in accuracy that triggers action? Regulators know you won't have a perfect answer. They want to see that you imagined failure before it arrived rather than discovering it after customers were harmed.

The Trap That Makes Good Governance Worse

Now the part that should genuinely worry you, because it's where the sophisticated organizations get caught.

I worked with a company that had every structural element of control in place: quarterly reviews, decision approval sign-offs, dedicated model risk managers. Then a regulator's hypothetical became real. A model recommendation produced a clearly unfavorable outcome for a member of a protected class, and the only question on the table was the one nobody had built for: why did the system score that customer that way?

They couldn't answer it. Not for want of a sign-off: the approval was on file, signed, dated. They couldn't answer it because nothing about *what the model had learned from* had been preserved. The training-data pipeline had been rebuilt twice since deployment with no snapshot of the earlier versions, so the dataset that produced that score no longer existed anywhere in the company. The feature-engineering logic lived in Jupyter notebooks belonging to a model owner who had moved to a different team. Nothing tied a scored customer back to the data that scored them. They had governance but no explainability, documentation but no understanding — and no way to establish whether the outcome was a defect, a data artifact, or exactly what the system was designed to do.

And here is the cruel twist: a governance structure built on that foundation is worse than none at all. It proves you had a process. It proves you had accountability. Which makes the unexplained bad outcome look *deliberate* rather than accidental. The paperwork that was supposed to protect you becomes the evidence against you.

This is why the first thing a regulator reaches for is not your framework. I asked someone in the policy space at a financial regulator what he'd look at first when he started examining banks' AI systems. He didn't say "your governance framework" or "your audit logs" or "your risk register." He said: "Show me the decisions this system is making. Now show me three examples of bad outcomes. Why did the system make those decisions, and why didn't anyone catch them?"

That is not a compliance question. It's a comprehension question. It reorders the entire priority stack. Before a regulator cares whether you have governance, they want to know whether you understand your own system well enough to explain its behavior. Governance documents are meant to be *evidence* of that understanding —

not a substitute for it. Most enterprises have built the evidence and skipped the understanding.

The Fix Is Decision Infrastructure

Here's the good news, and it's genuinely good: the fix is not technically sophisticated. It's not a better model or a cleverer interpretability method. It's decision infrastructure — the same dull process work banks have done for lending decisions for decades, pointed at a business process that happens to run on math.

Decision infrastructure has a small number of moving parts, and you can build all of them.

A **decision record** sits at the center: Chapter 1's template, filled in before the system goes live. What turns it from an internal artifact into a regulatory one is timing and coverage. Most organizations don't have this document for a single system; a regulator would expect it for every material one, dated before deployment, because a record written after the outcome is an explanation, and everyone has an explanation. If you're reading this with systems already in production, build the records backward: pick one deployed model, walk the decision, find out who chose it and what they rejected, and write down only what you can actually establish. A backward-built record with honest gaps is worth more in an exam than a clean one assembled from memory, and it shows you exactly where your governance was silent.

Around that record you keep an **audit trail** with reasoning attached. Not "we deployed version 2 on this date," but "we changed the monitoring thresholds because of this, here's who approved it, here's why." An **escalation record** captures not that nothing went wrong (something always goes wrong) but that you caught it and decided what to do about it. Most enterprises have an incident-response plan; almost none have a "what did we decide when we found the problem" plan, which is the exact thing a regulator wants to read. And **approval authority** rests with the named person Chapter 1 already put you on the hook for, with one regulatory qualification: what an examiner is tracing the reporting line *for* is P&L authority. Responsibility without a budget, and without the standing ability to pull the system, is not what a regulator means by an owner.

Then there's the functional half: the understanding the governance is supposed to evidence. You maintain genuine **explainability**: can you pull a specific decision and explain it to someone who isn't your data scientist? Not "the model scored it," but "the model weighted these factors, which led to this result, and here's why those weightings made sense for this use case." Some teams get there through SHAP values or other interpretability tooling; others keep clear decision rules. The method matters less than the commitment: you're building to *be* explainable, not building and hoping explanation shows up later when a regulator asks.

You monitor for the right kind of drift. Chapter 12 draws the line between system health and behavioral health; the regulatory point is narrower than the full taxonomy. Statistical drift is a performance problem, and you can fix it. **Behavioral drift** (decisions shifting in ways nobody designed for, one segment steadily getting different treatment) is the kind that ends in an enforcement action, because by the time it surfaces there is a population of customers who were treated differently and a monitoring record showing you were watching a metric that could never have caught it. Alert on the second one specifically, and keep the alert history, because "we monitor for this" is a claim and a log of alerts with dispositions is evidence.

You keep **decision memoranda**, one page per major choice: the choice, the alternatives, why you chose what you chose, and, critically, what would tell you that you were wrong. When something later fails, that page lets you distinguish "the system worked as designed but the design was wrong" from "the system didn't work as designed." That is a distinction most post-mortems can't make, and an expensive one to lack. And you invest in **traceability of inputs**, which is where nearly everyone is weakest: where did the training data come from, has it been modified, who had access, can you reconstruct the dataset from six months ago if an outcome looks discriminatory? That's part governance, mostly plumbing (data versioning, provenance, lineage), and there's no shortcut. It also happens to be the only way to audit your own bias honestly, because you cannot fix what you cannot trace.

If you want a single diagnostic to know where you stand, it's this: pick a model that's in production right now and explain, in plain language, why it made a specific decision in a specific case. If you can't, you don't have compliance readiness. You have documentation. Once you *can*, build the governance around the assumption that you can always be asked to explain (inherently explainable architecture, inherently traceable data, inherently defensible decisions), and the paperwork will follow honestly from work you've actually done.

None of it is impressive. But when a regulator asks for it, you have it, and that is the entire difference between a review that goes smoothly and one that doesn't.

Compliance Theater vs. the Real Thing

The failure mode has a specific shape in an exam room, and it is worth naming precisely, because it is seductive. Theater is when a regulator asks "did you think about this?" and you hand over a document that proves you have a process and demonstrates that you thought about nothing in particular. It is a real artifact, produced by real work, by people who were trying. It just answers a question nobody asked. Real compliance is when the same question gets "yes: here's what we thought, here's what we decided, here's how we're managing it," and there's a record behind every clause.

Which leads to the most counterintuitive thing I can tell you about regulatory conversations: admitting what you don't know is stronger than performing certainty. "We've deployed this system and we monitor X, Y and Z" is a strong answer. "We monitor what we can" is a weak one. And "we don't fully understand how this system behaves in edge cases, so we route those to human review and hold conservative thresholds until we do" is a genuinely *defensible* answer — a known unknown with a mitigation bolted to it, which is the exact shape of a decision someone made on purpose. What is not defensible is having deployed it, not understanding what it does, and having nobody watching. Examiners are not looking for certainty. They're looking for whether you can tell the difference between what you've established and what you've assumed.

The companies handling this well are building decision infrastructure, not governance infrastructure. They keep records of decisions. They're explicit about who decided what. They document their reasoning before the fact rather than reconstructing it after. It's the only thing that survives contact with a regulator who knows what to ask.

The Advantage You're Throwing Away

Everything above concerns the examiner. Chapter 20 takes up the harder audience, which is a court, and the same decision record is what answers both.

All of which points at something compliance functions rarely say, because it sounds like a slogan: none of this is a tax. The story most companies tell themselves is that compliance is friction: governance adds process, rules limit innovation, it's the price of operating in a regulated industry. That's backward. The companies getting it wrong aren't moving faster; they're moving blind, and speed only costs more when you hit the wall. The ones getting it right deploy systems that stay in production, make decisions that don't need to be remade, and can answer the question a board eventually asks, not "have we passed an audit," but "do we actually know we're compliant?"

Three shifts in language do most of the work here, and they're worth stealing verbatim. From "are we compliant?" to "what are we assuming?" Most compliance questions are risk questions wearing compliance language, and a regulator asking about explainability is asking what could fail that you haven't prepared for. From "tell me what we can't do" to "tell me what we're responsible for." Framed as restriction, compliance produces "we can't deploy until X passes Y"; framed as scope, it produces "if we deploy this, we own the consequences," which is where actual compliance lives. And from document collection to decision documentation: most enterprise compliance is archaeological, digging up evidence after the fact, when the only way to catch assumption drift while it's still cheap is to write down what you're deciding before you deploy. The practical consequence is a rule you can apply on Monday: a system that can't be governed clearly creates liability even when it creates business value. You're allowed to take that trade. You just have to take it consciously, with a name on it, rather than discovering later that you took it by default.

Regulatory frameworks for AI are still emerging, and that's the urgency. They will crystallize, and they will branch, jurisdiction by jurisdiction, philosophy by philosophy. When they do, the organizations that have been building decision accountability from the start will make an easy transition. The ones that built technical governance and hoped it would satisfy the regulatory need will be reconstructing a foundation while standing in an exam room. When the regulator says "show me your decision log," you're going to want to already have one.

The Bottom Line

Regulators aren't auditing your architecture; they're auditing your judgment. The violation is never "the model was imperfect." It's "you deployed it without knowing the risk, deciding it was acceptable, and being able to prove both." That means the highest-value work isn't more fairness metrics or fatter model documentation; it's boring decision infrastructure: a named accountable owner with real authority, a written decision record dated before deployment, real explainability, and monitoring for behavioral drift, not just statistical drift. Build that and you don't just pass exams; you move faster, because an organization that understands its own risk can act with confidence while its competitors move blind. Compliance done right isn't the tax you pay to deploy AI. It's the foundation that lets you deploy it and keep it running.

Field Kit

1. The exam rehearsal (run it before an examiner does). The five-part framework earlier in this chapter isn't only what a regulator walks in with — it's a drill you can run yourself, out loud, with the people who would actually be in the room. Who is accountable for this deployment, by name, and what authority do they hold? What risk assessment did we do *before* deployment, including which alternatives we evaluated and rejected? What specific threshold or observation would make us pull it? Who is genuinely watching, and how often? And can we explain one real decision this system made, in plain language, to someone who isn't a data scientist? Any answer that arrives as a document rather than a sentence is a governance gap, not a documentation gap — you can't write your way out of it after the fact.

2. The registry-vs.-log audit. When a regulator says "decision log," they do not mean your model registry, and the two get confused constantly because both are spreadsheets with system names in them. A registry answers *what is running*: name, version, owner, metrics, deploy date. A decision log answers *who decided to run it, knowing what*. Go down your AI inventory and mark which of the two you actually hold for each system. Wherever the honest answer is "registry" and the system is material, the missing artifact is the Chapter 1 decision record — dated before deployment, or built backward with its gaps left visible rather than papered over.

3. The readiness diagnostic. Pick a model that's in production right now. Pull one specific decision it made and explain, in plain language, why it made that decision. If you can't — because the owner moved teams, the pipeline changed, or the logic lives in an unmaintained notebook — you have documentation, not compliance readiness. Run this on your three highest-risk systems this quarter.

4. The theater-vs.-real test. For each governance artifact you own, ask: when a regulator asks "did you think about this?", does this artifact prove we *thought about* *something specific* — or does it only prove we *have a process*? Retire or rewrite anything that only proves process. A framework that shows accountability without showing reasoning makes a bad outcome look deliberate.

5. The behavioral-drift alert history. Audit your monitoring stack against Chapter 12's distinction: how many alerts watch accuracy, and how many watch whether specific segments are being treated differently or outcomes are diverging from design intent? The regulatory asymmetry is the point — statistical drift is a performance problem you fix quietly; behavioral drift arrives with a population of affected customers attached. Add at least one behavioral signal per protected or high-impact decision, and retain the alert history including the alerts that resolved as nothing, because that log is the only evidence that "we monitor for this" was ever true.

6. The P&L test on the named owner. Take the name Chapter 1's rule gives you for each material system and check three things against it: does this person control a budget, can they pull the system out of production without asking permission, and does the cost land on them if it fails? All three, or the name is decorative. This is precisely what an examiner traces reporting lines to find out — because "accountable" without authority is a job title, and a job title has never pulled a model out of production.

CHAPTER 20

Legal Exposure: Liability, IP & Copyright

Part VII · Compliance, Law & the Regulatory Future

A general counsel stopped a room with a single question.

The company was deploying a large language model as an internal knowledge assistant. It was a reasonable project, competently run. The team had thought through data security. They'd built audit trails. They'd set usage policies. By every measure a project review would use, they were ready.

Then the GC asked: "What happens if the model hallucinates and gives someone bad advice they act on?"

Not a weird question. A straightforward one. And it turned out nobody had a clear answer.

The conversation that followed was instructive. The data science team said: well, it's a language model, hallucinations are a known limitation. The product team said: we have a disclaimer. The legal team said: disclaimers don't protect us from liability.

And that's where it usually breaks down.

Here is the reframe this chapter is built on.

The myth: legal exposure arrives when an AI system breaks, and disclaimers plus a thick stack of controls are what protect you.

The reality: the worst exposure comes from systems working exactly as designed, and what protects you is a named person who consciously decided and a documented provenance trail for the training data.

Most organizations understand, by now, that AI systems can fail. What most are still developing is any real understanding of the *structure* of liability when they do. And here is the uncomfortable premise of this chapter: the failures that create the worst legal exposure are usually not the ones where the system breaks. They're the ones where the system works exactly as designed — and confidently tells someone something false, or makes a decision that, when you go looking for who made it, nobody actually made. Underneath both sits a second, quieter exposure: the training data your models are built on, which you probably don't fully own and have never verified.

The scariest liability isn't the moment someone is wronged and sues you. It's the moment the harm is obvious, the accountability is clear on paper, and there is functionally nothing to sue. This chapter is about that, and about the IP problem you've been compounding the whole time, and about the work that actually protects you from them.

When the System Works and Still Creates Liability

Start with hallucination, because it's the failure mode most organizations have never built legal thinking around.

For decades, organizations have built legal frameworks for system failures they understand: databases go down, processes fail, data gets lost. There are precedents. There are insurance products. But a system that confidently tells someone something false is a newer problem, and the legal structures around it are still settling.

The risk isn't uniform. It depends entirely on the path from output to consequence.

If your system is purely informational (you're using it to summarize documents internally), the risk profile is one thing. Someone reads the summary, notices it's wrong, corrects it. Inconvenient. Not necessarily a legal problem. But if your system is integrated into a workflow where a human is *supposed* to verify the output, and that human reasonably relies on the system's apparent confidence, you have a different risk structure entirely. And if your system is trained on your proprietary knowledge and someone acts on its hallucinatory output in a way that damages a customer relationship, a business deal, or a compliance outcome, you have a third.

None of these are hypothetical. "The model told our sales team that we offered a service we don't actually provide." "The system generated a draft contract clause that contradicted our terms, and the team didn't catch it." "The model synthesized customer data in a way that suggested we had capabilities we don't have." None of these broke the system. All of them created liability. The system did precisely what it was built to do (generate a confident, fluent output), and the confidence was the problem.

This is why the disclaimer instinct is wrong, and why the GC in that room was right. A disclaimer manages expectations. It doesn't reallocate the exposure that lives in the gap between what the system said and what someone did about it.

What you actually need before any system with meaningful hallucination risk goes from pilot to production is a document. Not model documentation. Not a risk register. Not a binder. One page, written by legal, not by data science, that answers a single question: *if this system confidently says something false, and someone acts on it, what is our exposure?*

Call it a hallucination liability assessment. To write it, legal has to work through five things:

- **Context.** What decisions or actions does this system enable or inform? What's the chain from system output, to human decision, to real-world consequence?
- **Confidence.** When this system is wrong, is it *obviously* wrong? Or does it present falsehoods with the same fluent confidence as facts? Overconfident falsehoods create liability that visibly uncertain outputs don't.
- **Verification.** Is there a verification step in the workflow? And, critically, is it designed to catch hallucinations, or just to confirm that the system ran?
- **Scope.** Who can see the output? Internal team, or customers and partners? Scope multiplies liability.
- **Exposure.** If the system hallucinates and someone acts on it, what's the worst-case consequence? A client relationship? A compliance violation? A wrong regulatory filing? Put a dollar figure on it. Not a severity rating — a number, with the arithmetic shown. This is what determines how much you need to mitigate.

That last one is where most assessments go soft, so it's worth seeing one filled in. A team running this on an internal sales-support assistant wrote the Exposure line like this: *worst case, the assistant states a service commitment we don't offer, a rep repeats it in writing, and we either deliver at a loss or walk it back. Four accounts in the pipeline where that's live, combined ARR $4.2M. Cost to honor the misstated scope: roughly $600K per account. Odds we catch it before it goes out in writing: call it even.* Two of those numbers are close to guesses. It didn't matter. The estimate is what turned "hallucinations are a known limitation" into a decision: a second pair of eyes on anything naming a deliverable, about ninety minutes a week of one person's time. Ninety minutes against $4.2M is not a hard trade. It only becomes one when nobody has written the $4.2M down.

Once you've answered those, you're finally in a position to decide whether to deploy as-is or change how the system is used. Maybe you add human verification at a specific step. Maybe you constrain who can see the output. Maybe you write disclaimers that mean something instead of disclaimers that perform diligence. Maybe you don't deploy it the way you first planned.

Most organizations skip all of this. They get to deployment, ask one question (*does it work?*), and if it works, they ship. Then they're surprised when the liability question surfaces in month four because someone acted on something the system made up. Legal is increasingly asking for the assessment, but asking late, trying to bolt guardrails onto a system designed without a single legal thought in the room. That's expensive and slow. The companies handling this well aren't the ones with perfect models. They're the ones clear about the risk and clear about what they did to manage it. That clarity starts with a document legal owns.

The Liability With Nothing to Sue

The hallucination problem is real, but it's the one your legal team can at least imagine. There's a worse one they've probably built a framework around without ever confronting: the scenario where the liability is clear and there's nothing to sue.

Here's the pattern. Your organization deploys an AI system. It makes a bad decision. Someone is harmed. And the decision was, technically, made by the AI system — which is a tool.

Your company says: "We didn't make the decision. The model did. We relied on the model's recommendation because it had been validated and was operating as designed." The person who was harmed says: "I don't care about your controls. I was denied credit, or fired, or denied medical treatment. Someone should be accountable." And when everyone goes looking for who made the decision, the honest answer is usually: nobody. The decision happened. Nobody made it. Nobody can be held accountable.

This is not negligence liability, where you could argue the organization was careless. This is *structural* liability, where the organization constructed a system in which decisions happen without anyone deciding.

You might think that's good news. If nobody made the decision, nobody can be held liable. That is exactly backwards, and it's the most expensive misreading in this whole area. The emerging position among courts and regulators is the opposite: if you've deployed a system that makes decisions affecting people, *someone* at your organization bears responsibility for those decisions. The problem is that companies have distributed responsibility so widely that it disappears: the escrow problem from Chapter 3, seen from the legal side. Where Chapter 3 counted the cost in decision quality, the legal consequence is narrower and worse: escrow manufactures liability with nobody attached to it. It leaves a permanent record in which the review either happened or it didn't, and everyone nominally responsible can point, accurately, at someone else. That record is what gets read back to you in a deposition.

Watch it happen in a single role. A lending officer reviews a model's recommendation on a loan that gets denied. On paper this is the control working as designed: a trained human with delegated authority looking at the case before anything happens to the applicant. In practice the officer clears north of eighty files a shift, the model agrees with the eventual outcome nearly every time, and the review compresses into a confirmation. The officer hasn't made a decision; they've deferred to one. When the denial is challenged eighteen months later, you cannot hold them accountable for a decision they never really made: the accountability your org chart promised was never there. It's the loan model from Chapter 2, reviewed by everyone and decided by no one, except spread across ten thousand files rather than concentrated at one gate. Diffusion at the gate and diffusion at the point of use produce the identical artifact: a harmed person, a clear harm, and no author.

Regulators have noticed, and their question is changing. It used to be "who made the bad decision?" Now it's closer to "why did your system make a bad decision without anyone noticing?" That's individual accountability giving way to system accountability. Organizations that have been through a regulatory review tend to discover the same painful exchange. The regulator asks: "Your lending model denied this application. Walk me through how someone with decision authority reviewed this recommendation." The organization answers: "Our model is trained to be reliable. We monitor it quarterly. If it falls below performance thresholds, we retrain." The regulator pushes back: "That's not what I asked. Someone made this specific decision. Tell me who. Tell me what they knew. Tell me what decision authority they had." And the organization can't answer, because nobody actually had authority for that specific decision. The model recommended it, it fell inside the parameters that don't trigger escalation, and so the decision happened without anyone deciding.

The regulatory position, stated plainly, is: you can't build a system where decisions happen without someone deciding. Either the model has authority, in which case you'd better be able to defend the model's decision, or a person has authority, in which case you need to show they exercised it.

This is where traditional frameworks fail. Most companies approach AI liability through the lens of product liability: you build something, it fails, you're liable if you were negligent, and your defense is "we were careful, we tested it, we built reasonable controls." That works for a car with a brake failure. It does not work for AI, because the question isn't only whether the system was tested: it's whether anyone was actually responsible for *using* it. Organizations almost always have the product-liability half: testing, monitoring, control standards. They almost never have the decision-authority half: who decides what, under what authority, with what information.

So what actually protects you is the opposite of what your instincts suggest. It isn't "we have no responsibility because the model decided." It's "one identifiable person in this organization had clear authority and responsibility, and made a conscious decision." You are more protected, legally, by one person who says *we're going to use this model, and here's why I'm comfortable with that* than by a wall of elaborate controls with nobody clearly deciding. The controls are good. They just don't protect you if they're controls without an owner.

In practice, the organizations that get this right do four things. They make decision authority explicit. And the legal wrinkle goes past naming an owner: that owner has to have been in a position to *actually decide*, with the information, the time, and the standing to say no. An owner who could not realistically have chosen otherwise (the lending officer with eighty files a shift) gives liability nowhere to land. They document the conscious choice: what that person knew about the model, what risks they were accepting, what monitoring they'd put in place. They hold that person accountable for the decision. The question, if something goes wrong, is "was this decision reasonable given what was known at the time?" rather than "did the model fail?" And they design escalation so that when the model produces an unexpected outcome, someone with authority reviews it and decides whether the original decision still stands.

The protection comes from being able to show that someone thought about the risks and chose to proceed anyway — not from showing you had no responsibility. Which means your governance structure has to be read as a liability instrument, not just a control instrument. If it's a matrix of approvals where nobody clearly decides, you're building liability, not protecting against it.

The Data You Don't Fully Own

Now the second axis. Your organization is probably training a model on data you don't fully own, and you haven't thought hard about what that means.

This isn't theoretical. It's becoming the reason some organizations can't deploy models they spent months building, why legal is suddenly in the room for product decisions, and why audits are unraveling. The legal landscape around training data is messy enough that most teams are operating on assumptions, so it's worth separating myth from reality.

"If it's publicly available, we can use it for training." Publicly available is not the same as publicly licensed for machine learning. A dataset scraped from websites is not a dataset licensed under a Creative Commons agreement. A corpus of text published on the internet is not automatically yours to train a foundation model on. Copyright holders have claims on their work even when it's freely accessible; some have already started suing, and the ones who haven't still have standing to. You can train a model on public data. You cannot assume you own the right to deploy it without knowing where that data came from.

"Fair use lets us use any training data we want." Fair use is a legal defense, not a permission. It's evaluated case by case on four factors: the purpose of the use, the nature of the work, the amount and substantiality used, and the market impact. Training a commercial model is not obviously fair use: the amount used is massive (you're using the entire work), and the market impact could be significant (you may be competing with the original creator). Courts might find fair use applies. They might not. "Hoping you're right" is not a risk framework. Treat fair use as a defense you might raise if you're sued, not a license to use whatever you want.

"Small amounts of copyrighted data in a huge training set are negligible." There's no legal threshold where a small amount of copyrighted material becomes acceptable. The question isn't "how much?" It's "did you use copyrighted material without

permission?" You can't solve an attribution problem by dilution. A tiny amount of someone's work without permission isn't better than a large amount: it's still using it without permission. And the arithmetic runs hard against you, because US statutory damages are assessed *per work,* not per corpus: $750 to $30,000 for each infringed work, up to $150,000 where infringement is found willful. Say two percent of a 40,000-document internal corpus turns out to be material you had no clean right to train on. Two percent sounds like noise. It's 800 works: $600,000 at the floor, roughly $24 million in the ordinary range, nine figures if a court decides you should have known. The share of your training set is the wrong denominator; the count of works is the exposure.

"Synthetic data solves the problem." Sometimes. Generated from scratch, you own it. But many approaches seed generation with real data, and if you produced synthetic data by fine-tuning a model trained on licensed data, you've inherited the licensing problem. Synthetic data is useful. It is not an automatic escape hatch.

What makes this genuinely hard is the layer underneath: the foundation models you're fine-tuning were themselves trained on data whose provenance is unclear, at internet scale, under active legal scrutiny. Fine-tuning on your own data creates derivative works of derivative works — your risk layered on theirs.

And here is the blind spot that catches most enterprises. The typical audit asks: "Are we using a licensed model or an open-source model?" If it's licensed, they assume the vendor handled copyright. If it's open-source, they check the license (GPL, MIT, Apache) and call it secure. What they miss is the training data itself. When a foundation-model company faces copyright liability, they shoulder it: they chose what to include. But when *you* build an internal system (a support chatbot trained on your documentation, a code tool trained on your private repositories, a classifier trained on labeled examples), you are the one making those decisions and controlling that dataset. A fully licensed base model fed proprietary customer data you never had consent to train on doesn't make the base model's license the story, and an open-source model trained on data you don't have clean rights to doesn't make the model's license your primary risk.

Three litigation tracks are running at once. The first is generative-AI companies and the datasets they trained on: the high-profile cases alleging that training on copyrighted material without permission or compensation violates copyright. The second is enterprise use of foundation models to generate content that replaces work that would otherwise have been licensed. The third (less visible, and arguably the most dangerous for enterprises) is the training data you're collecting and using *internally.* Most enterprises aren't thinking about the third category. It's the one they most directly control.

Two things are changing fast. First, the theories of liability are starting to flow backward, from end users toward the data sources and the choices made about what data to use. The comfortable assumption that copyright enforcement was the model builder's problem, not the model user's, is breaking down. Second, "we didn't know it was copyrighted" is not a defense if you had reason to know; and a large enterprise has more reason to know than a startup. Courts will assume you had the capacity to verify your sources. The framing is moving from "did you train on copyrighted material" to "did you have a process for knowing what you were training on." That shift matters more than it looks: process failures are easier to prosecute than content mistakes. Your model might be excellent and the data might be exactly right for the problem, but if you can't defend your rights to use it, you can't deploy it. That's not a legal problem you solve later. It's a governance problem you solve now.

What Actually Protects You

Notice what both halves of this chapter converge on. The liability problem and the IP problem look unrelated (one is about decisions, the other about data), but they resolve to the same discipline. In both, the thing that protects you is a named human owning a conscious, documented decision *before* the system goes live, and the thing that sinks you is a diffuse process where everyone contributed and no one decided.

On the IP side, that discipline is provenance. Know what data you're training on: an actual list, sources documented, licensing and ownership metadata captured alongside. For each source, verify the rights: not "is it publicly accessible?" but "do we have the rights to train a commercial model on it?" That means reading the license agreements and, when licensing is unclear, writing the uncertainty down rather than papering over it. Keep the due-diligence records, because willful infringement carries five times the statutory ceiling of the inadvertent kind, and a documented good-faith effort is most of what moves you between the two. Revisit vendor contracts: the boilerplate you signed three years ago almost certainly doesn't address model training, and licensed customer or third-party data should now explicitly cover, or explicitly prohibit, training use. Then make provenance a governance gate, the same way you gate on model validation: for any model going to production, legal signs off on the training data, or there's a documented risk assessment saying why not.

On the liability side, that discipline is decision authority. For every system in production, you should be able to answer three questions cleanly: Who has the authority to decide whether we deploy this? What did they know about the model's risks when they decided? If something goes wrong, who reviews that outcome and decides what to do? If you can't answer those, you have a liability problem, not because your model is bad, but because nobody decided to use the model. Fix that before you invest in more controls or a more comprehensive framework. More controls without an owner is more of the exact thing that dissolves accountability.

The reason so few organizations do either well is that there's no demo in it. Price it honestly: legal's exposure page for one production system is a morning's work; a provenance sheet for a fine-tuning corpus is an afternoon if the sourcing was disciplined and a couple of weeks if it wasn't. Set that against 800 works of per-work statutory exposure, or a regulatory review where you can't name who decided, and it's the cheapest insurance on the roster by orders of magnitude. Which is why it keeps losing the argument for headcount: nobody funds insurance against a number nobody has written down. Early movers get a real advantage: they know what they can and can't do, and they hold a documented process that demonstrates reasonable care. Everyone else has a much harder conversation mid-deployment, when they discover their training data has clean-title problems, or that the decision that harmed someone was one nobody actually made. The copyright landscape and the liability landscape are both clarifying month by month. The visibility organizations have into their own systems is not. That gap is where the exposure sits.

The Bottom Line

The legal exposure that should worry you most isn't the dramatic failure that ends in a lawsuit. It's the quiet kind: a decision that harmed someone but that nobody, on inspection, actually made, and a stack of training data you can't prove you had the rights to use. Disclaimers, elaborate controls, and "the model decided" are not protection; in the accountability case they are the source of the problem. What protects you is human and specific: one named person who was in a position to decide, decided consciously, and documented why, a real verification path for confident falsehoods, and a documented provenance trail for every dataset your models learned from. Courts and regulators are moving from "did the system fail?" to "who was responsible for using it?" and from "did you train on copyrighted material?" to "did you have a process for knowing what you were training on?" In both, process and clear ownership are what you'll be judged on. Do that work before deployment, or inherit the far more expensive version of it later.

Field Kit

1. The pre-deployment liability question set (bring to the go-live meeting). Before any system with hallucination risk moves from pilot to production, make legal answer, on one page: *Context* — what's the chain from output to human decision to real-world consequence? *Confidence* — when it's wrong, is it obviously wrong, or fluently wrong? *Verification* — is there a step designed to catch hallucinations, not just confirm the system ran? *Scope* — who sees the output, and how far does it travel? *Exposure* — the worst case as a dollar figure, with the arithmetic shown?

2. The decision-authority test (run per production system). Answer cleanly, or you have a liability problem: Who has the authority to decide we deploy this? What did they know about the model's risks when they decided? If something goes wrong, who reviews the outcome and decides what to do? Legal's addition to the single-owner rule: the owner must have been in a position to *actually decide* — information, time, standing to say no — or liability has nowhere to land.

3. The training-data provenance checklist (per dataset). Where did this data come from? Who originally created it? What license or terms govern its use — and do they cover commercial model training specifically? Was it collected with consent for *this* purpose? Did we verify any of the above, and where is that verification recorded? "Publicly available" is not an answer to the license question.

4. The myth-buster card (tape it to the data-sourcing decision). Publicly available ≠ licensed for ML. Fair use is a defense, not a permission. Damages run per work, so dilution is not a defense. Synthetic data seeded from real data inherits real-data problems. Open-source *model* license says nothing about your *training-data* rights.

5. The two lines legal owns on the decision record. The Chapter 1 decision record already carries the owner, what's being accepted, the thresholds, the monitors, and the escalation path. For any system with hallucination or training-data exposure, two entries on it belong to legal rather than data science: legal *writes* the exposure page from item 1 and attaches it, and legal signs off on training-data provenance — or files the risk assessment saying why you're shipping without that sign-off. The delta is authorship, not review; only the first reads as reasonable care. If neither line can be signed, the system isn't ready to deploy.

CHAPTER 21

Compliance That Survives Uncertainty

Part VII · Compliance, Law & the Regulatory Future

I spent a morning last year in a windowless room with the model risk team at a mid-size bank, going through what everyone had expected to be a victory lap. They had built an AI system for a high-stakes lending decision, run it through a fairness assessment, tuned it until the numbers cleared, and deployed it with a clean conscience and a clean audit trail. Their primary regulator's fairness requirement: met, documented, signed off.

Then a second regulator arrived with a second definition of fairness — harder to satisfy, focused on different demographic groups, measuring the disparity a different way. The system that had passed one exam failed the next, not because the system had changed, but because the meaning of the word "fair" had.

So they built a second fairness pipeline. Halfway through, someone re-read the first regulator's original guidance and found that it had asked for something close to the second definition all along — worded differently, buried in a different section, missed on the first pass. By the time I was in the room, they were maintaining three approaches to the same problem, because three authorities each wanted something slightly different and none of them would say precisely what. Their head of model risk put it better than I could: "We're not behind on the rules. We're behind on the rate the rules change."

That sentence is the whole chapter. Financial institutions are living it right now, and it previews what every enterprise operating across jurisdictions is about to face with AI. The problem is not that you don't yet understand the rules. It's that the rules will arrive fragmented, contradictory, and expensive to navigate. And the compliance program you're building to chase them is the wrong program.

Here is the reframe this chapter is built on.

The myth: regulatory clarity is coming, and when it arrives you'll know what to build.

The reality: clarity is coming, but it will land in pieces that don't fit together, and the only thing worth building now is the capability to flex when they change under you.

Most enterprises are waiting for clarity before they act. They're wrong, and the mistake is subtle, because waiting feels prudent. But the preparation that matters isn't "understand the regulations." It's "build the capability to survive regulatory fragmentation." Those are different projects, and only one of them survives contact with reality.

Different Rooms, Different Rules

Start with why AI is different from the compliance problems you've already learned to manage.

Regulatory divergence is not new. You've seen it in data privacy, where GDPR sits alongside a patchwork of US state laws alongside jurisdictions with no meaningful regulation at all. You've seen it in financial services, where capital requirements and lending rules vary by country. You've seen it in environmental compliance. And in each of those domains, organizations eventually learned to operate, usually by consolidating around the strictest requirement and applying it everywhere.

AI breaks that pattern, and it breaks it before the pattern has a chance to form. Privacy took two decades to fragment and then converge, and by the end there was a recognizable strictest requirement to consolidate around. AI is fragmenting while the first frameworks are still being drafted, and the pieces aren't the same shape: the EU's "high-risk system" and a US banking supervisor's "model risk" are not the same object, so building to one doesn't get you most of the way to the other. The rules are less settled inside each jurisdiction than they look from outside, too. Guidance gets reinterpreted between exam cycles, and documentation that passed last year comes back as a finding. And the per-jurisdiction cost is far higher than it was for privacy, because what regulators want here are engineering artifacts, not policy documents: fairness measurement, explanation, monitoring evidence, decision records.

Put those together and you get something the older domains took decades to produce: regulatory arbitrage. Global enterprises can effectively choose which rules to follow based on where they run their systems, where they process their data, and which governance structures they invest in. That choice has a material impact on both their risk profile and their competitive position.

Look at the actual landscape and the incentive becomes obvious. The EU is implementing the AI Act: risk tiers, requirements for high-risk systems, documentation mandates, audit trails. Comprehensive, and coming. The US has sector-specific guidance (the SEC on AI risk disclosure, the OCC on model risk management, the FTC on bias), but no unified framework: fragmented, and softer. The UK is principles-based: follow these principles, but we won't prescribe how. Softer still. China regulates content and outputs rather than governance, a different problem entirely.

Every enterprise I've worked with operating in more than two of those has a version of the same wall chart: jurisdictions across the top, requirements down the side. Every one has the same defect: the columns don't reconcile. There is no row you can satisfy once.

So a temptation appears. Invest heavily in EU-grade governance if you primarily serve EU customers, because you have to. Invest lightly in the US, because you don't have to and the rules are unclear. Run the same system differently in different regions, or route work to wherever the governance bar is lowest. Nobody makes this decision consciously — that's the point. But the incentives all point the same direction, toward the floor.

The Three Problems Arbitrage Creates

That drift toward the floor produces three problems, and they compound.

The first is **governance inconsistency**. A multinational can end up with two identical AI systems doing identical jobs in two different regions, governed completely differently. One has quarterly bias audits and an explicit decision-appeals process. The other has neither. Why? Because it can. This is unsustainable both as an ethical matter and as a legal one. Eventually one of the looser regions has an incident, and the question you'll be asked is not "did you meet the local requirement?" It's "why didn't you use the governance structure you were already using everywhere else?" The moment you can do something well in one place, your failure to do it in another stops looking like a resource constraint and starts looking like a choice.

The second is **incentive misalignment**. If operating a high-risk AI system costs meaningfully more in Region B than Region A, decisions will migrate to Region A. Not by decree — by gravity. If your customer base spans regions and your finance team reports that "governing this system for the EU costs 40% more," your product team will find perfectly reasonable-sounding grounds to process more decisions where governance is cheaper. I have never seen anyone write that decision down: no memo, no owner, just a routing rule and a quarterly cost target. The CFO isn't trying to weaken your controls. The incentive is simply doing what incentives do.

The third is a **coordination problem**, and it's the one that will grind at you competitively. Say you're a financial services firm operating globally. GDPR required privacy-by-design, so you built it. The EU AI Act requires transparency and explainability, so you're building that too. Meanwhile a competitor in a lightly regulated market is building none of it. They ship faster and carry lower compliance costs, and that is a real advantage. The only way to level the field is for everyone to play by the same rules, which is exactly what isn't happening.

There are really only two ways this ends. In the first, global enterprises consolidate around the strictest requirement and apply it everywhere, the way they did with privacy and financial regulation; that outcome is stable. In the second, fragmentation persists: different enterprises follow different rules, and the whole market becomes harder and more expensive to coordinate; that outcome is expensive and unstable. Between here and either endpoint sits a transition period, possibly a long one, in which different organizations make different governance bets based on where they operate, and the looser jurisdictions either tighten (expensive, and usually late) or the strict jurisdictions' rules leak out anyway through litigation, customer pressure, and reputational risk. You are living in that transition period right now.

The Wrong Question

So what do you do? Most enterprises reach for the intuitive answer, and the intuitive answer is a trap, because it starts with the wrong question: *Which regulation applies to us?* Answer that, the thinking goes, understand all the rules, then build to the strictest one. It sounds rigorous, exactly the kind of thing a serious compliance function should do. And in a world where the rules held still, it might even work.

The rules do not hold still, and "build to the strictest" fails for three concrete reasons. New regulations will keep coming, so you can never be built to all of them at once:

the target set is still growing. Regulators interpret their own rules differently over time; what counts as "fairness" or "transparency" or "accountability" is not a fixed quantity you can engineer against once and be done. And compliance cost scales with the number of requirements you're actively navigating, so building to the strictest interpretation of every rule simultaneously doesn't make you safe — it makes you uncompetitive, and it still doesn't cover the requirement that shows up next quarter.

This is exactly the trap the bank fell into. It built to a fairness requirement, then a second definition arrived, then it was maintaining three, and every one of those three was, at the moment it was built, the strictest interpretation anyone in the room knew about. "Build to the strictest" is not a stable position. It's a snapshot of a moving picture, and the moment you finish building it, it's out of date.

The right question is different: *How do we build adaptability into our AI systems so we can respond to multiple, changing, contradictory regulatory requirements without re-architecting every time?* That question doesn't ask what the rules are. It asks what capability you need so that the rules can change and you can absorb it. Stop trying to guess the destination. Build the thing that lets you keep moving.

What Adaptability Actually Looks Like

Here is where the work gets unglamorous, which is exactly why most organizations skip it. Adaptability isn't a policy or a principle or a slide. It's plumbing. The enterprises that survive regulatory fragmentation build three specific things.

Modular governance. Your governance framework should be separable from your systems. This sounds simple and it is not, because most enterprises have done the opposite: they've baked governance into the systems themselves. Decision logic is welded into model architecture. Monitoring lives inside inference pipelines. Audit trails sit in system-specific databases that were never designed to be queried by anyone but the engineers who built them. So when a regulator says "measure fairness this way," you can't just add a measurement. You have to re-architect the system to add it.

Modular governance means your monitoring, fairness assessment, bias detection, audit trails, and decision recording all happen in a governance layer that sits separate from the system layer. When a regulator asks for something new, you add it to the governance layer and leave your systems alone. This is genuinely expensive to build.

It is far cheaper than rebuilding everything, every time, for the rest of your operating life. That is the trade, and it is the whole game: pay for the seam now, or pay to tear the thing apart later.

Configurable processes. Your governance processes should not be hardcoded for one regulatory environment. One regulator requires that model decisions be human-reviewed before they touch a customer. Another cares less about pre-review and more about post-hoc audit. Your process should not be "human-review everything" or "review nothing." It should be: *we can require human review for certain decision types, in certain jurisdictions, in certain contexts; when a regulator changes, we reconfigure the process instead of re-engineering the system.* That is a design requirement, and you have to build it in deliberately. Configuration, not hardcoding. The difference between the two is the difference between a settings change and a six-month project.

Regulatory intelligence infrastructure. You need someone watching the regulatory landscape and translating it into implications for your systems, and it cannot be one person doing it in the margins of another job. It's a function. It tracks what each regulator is doing, how the text is being interpreted in exams, and where that leaves you. For every material change, it asks the same short list of questions: Does this apply to us? What do we need to do? How long will it take? What does it cost? What's the regulatory risk if we don't? And then it feeds the answers back into your governance roadmap, so that regulatory change enters your organization as planned work rather than as a fire.

The organizations ahead on this didn't start when the AI Act text landed. They started a build or two earlier, when someone at a whiteboard asked whether the fairness metric should live inside the model service or beside it. Answered one way, that question costs a sprint; answered the other, a year — and you don't find out which you chose until a regulator asks for a metric you don't have.

The Practical Path

You don't have to boil the ocean, and you shouldn't try. Adaptability is architectural work spanning your systems, your governance infrastructure, and your processes. But you can sequence it, and the sequence starts with one question: *What are the regulatory dimensions we're most uncertain about?*

For most enterprises, the answer comes down to four. **Fairness**: what counts as fair, and who decides. **Transparency**: what actually explains a decision. **Accountability**: who is responsible when it goes wrong. And **audit rights**: what a regulator can demand to see, and in what form. These are the dimensions where interpretations diverge most, which makes them the dimensions where flexibility pays off most.

For each one, the goal is optionality rather than a single frozen answer. On fairness, can you measure it multiple ways and report whichever metric a given regulator cares about? On transparency, can you explain a decision multiple ways to different audiences: a customer, an auditor, a court? On accountability, are your decision records flexible enough to map onto different accountability frameworks, or are they locked to one? On audit, can you hand different data and documentation formats to different auditors without a data-engineering project each time? None of these asks you to predict the rule. Each asks whether you can produce whatever the rule turns out to require.

This is not one project, and pretending it is one project is how it dies in committee. It's a program. But it is cheaper, reliably and measurably cheaper, than the alternative, which is being forced to re-architect every time a regulator changes their mind.

The Timeline, and the Test

You do not have years. The EU AI Act is going into effect. UK frameworks are tightening now. US sector-specific rules are moving. Financial regulators are already fast and getting faster. The window in which "we're waiting for clarity" is a defensible answer is closing, and in some sectors it has already closed.

Nothing about that is hypothetical, and the clearest illustration arrived while this book was in production.

The EU AI Act set 2 August 2026 as the date its high-risk obligations would apply. Compliance programs across every multinational I worked with in 2025 were built against that date. Then, on 24 July 2026, Regulation (EU) 2026/1744 — the Digital Omnibus on AI — was published in the Official Journal, entering into force three days later. It moved the high-risk obligations for stand-alone systems to 2 December 2027, and for AI embedded in regulated products to 2 August 2028. The stated reason was not a change of policy. It was that the machinery the rules depended on, the national authorities and the harmonized standards, was not ready.

Read what that actually did to a governance system. The headline deadline slipped sixteen months, six days before it was due, and entry into force was compressed to three days because the old date was arriving. Meanwhile the transparency obligations in the same statute did not move at all, and a new prohibition was added on a use case nobody had been tracking. So the correct response was not “we have more time.” It was: three obligations moved, one didn’t, one is new, and every one of them has a different date.

Organizations that had written 2 August 2026 into their control descriptions, their audit calendars, their vendor contracts, and their board reporting spent the weeks that followed re-planning. Organizations that had built dates, scope, and jurisdiction as configuration changed the configuration. Same regulation, same news, two completely different quarters. That is the entire argument of this chapter, delivered by a regulator, on a schedule nobody chose.

There is a single diagnostic that cuts through all of this, and I ask it in governance reviews because of how reliably it changes the temperature of the room: *When a regulator changes their mind, can we respond in weeks or months, or do we need six months and a re-architecture?* The useful part isn’t the answer but who believes it. At the bank, the engineering lead said four weeks and the head of model risk said two quarters, and the gap between those numbers was the finding.

If the honest answer is six months, you are already exposed. Not exposed someday, when a rule lands — exposed now, because the response time is the vulnerability. A regulator doesn’t need to catch you non-compliant to hurt you. They only need to change a requirement faster than you can absorb it, and in a fragmented landscape, several of them will, at once, in different directions. The enterprises navigating this well are not the ones who guessed the rules correctly. They’re the ones who built so that guessing wrong is survivable.

There’s a comforting story that the organizations ahead today got there through foresight. Mostly they got there through luck: they made governance choices years ago, before the rules were clear, and those choices happened to line up with what regulators now want. Luck is not a strategy, and it doesn’t renew. The organizations that will be ahead in five years are the ones assuming their luck runs out and building accordingly: choosing a governance standard deliberately, documenting every place they deviate from it and why, and investing in the capability to reconfigure rather than rebuild.

Which brings it back to the choice underneath the whole chapter. If you run different governance in different regions, that can be a perfectly defensible decision: *"we use lighter governance in Region A because the regulation there is lighter"* is a sentence you can say to a regulator and survive. *"We accidentally built different systems and never noticed"* is not. The difference between those two sentences is not the systems. It's whether you decided on purpose and wrote down why.

The enterprises that build adaptability now will spend money. The enterprises that don't will spend more money later — and they'll spend it under pressure, in a crisis, with a regulator already in the room. Both of them pay. Only one of them chooses when.

The Bottom Line

Regulatory clarity for AI is coming, but it will arrive fragmented, contradictory, and expensive to navigate, and no two frameworks will fully align. Chasing today's rules (understanding them all and building to the strictest) is a snapshot of a moving picture that's out of date the moment you finish it, and it pulls your systems toward the lowest governance floor through arbitrage you never consciously chose. The durable move is to stop asking "which regulation applies?" and start asking "how do we adapt without re-architecting?" That means building three things: modular governance separable from your systems, configurable processes you can reconfigure instead of re-engineer, and a standing regulatory-intelligence function that turns regulatory change into planned work. The test is whether you can respond to a changed requirement in weeks, not six months; if it's six months, you're already exposed. You will pay either way. The only question is whether you pay now, on your schedule, or later, in a crisis, with a regulator watching.

Field Kit

1. The governance-review question set. Bring these five questions to your next AI governance or risk review and require real answers, not reassurances: - When a regulator changes a requirement, can we respond in weeks/months, or do we need six months and a re-architecture? - Do we have two systems doing the same job governed differently across regions — and if so, did we decide that on purpose and document why? - For fairness, transparency, accountability, and audit — can we produce more than one answer, on demand, for different regulators? - Is our governance built *into* our systems, or does it sit in a separable layer we can extend without touching the systems? - Who owns regulatory intelligence as a named function — and if the answer is "nobody, really," when does that change?

2. The exposure ranking, worst first. Don't score your systems — rank them. List every material AI system and sort the list with three questions, applied in order: Is governance separable, or welded into the system? Would a new requirement land as a configuration change or an engineering project? Is anyone watching the frontier for the dimension this system is most exposed on? Welded-in governance ranks above governance that's merely awkward to reach; a system nobody is watching ranks above one someone is. Argue the order out loud with engineering and compliance in the room — the disagreements teach you more than the ranking does. The top of that list is your roadmap, and its top entries are the re-architectures you'll do under pressure if you don't schedule them first.

3. The four-dimension flexibility checklist. For each high-risk system, confirm you can, on demand: measure fairness multiple ways and report the metric a given regulator wants; explain a decision differently to a customer, an auditor, and a court; map your decision records onto more than one accountability framework; and export documentation in different formats for different auditors. Every "no" is a future emergency project.

4. The deviation register (one-page template). A living log with four columns: *System / Region, How its governance differs from our global standard, Why (the deliberate reason), Approved by / date.* If a region's governance differs and there's no row for it, you don't have a defensible decision — you have an accidental policy gap. Fill it in before a regulator asks you to.

5. The regulatory-intelligence intake. A standing five-question form the intelligence function completes for every material regulatory change: Does this apply to us? What do we need to do? How long? What does it cost? What's the regulatory risk if we don't? Route every completed form into the governance roadmap as scheduled work, so change arrives as a plan, not a fire.

PART VIII

From Pieces to System

CHAPTER 22

The Systems View: Fifteen Pillars as One Operating System

Part VIII · From Pieces to System

I sat through a readout at a mid-size property insurer that had done everything the textbook told it to do. It had stood up a model-risk framework, and the framework said the sensible thing: high-stakes models get a retraining audit every quarter. It had stood up a change-management process, and that process said the equally sensible thing: no model change ships without two weeks of testing. It had signed with a platform vendor whose model versions refreshed on a weekly cadence, because that vendor was the best on the shortlist. And it had built a cost-control regime that budgeted around inference volume, because inference volume is what shows up on the invoice.

Every one of those decisions was defensible on its own. An auditor would have approved each in isolation. Together, they described a rule that could not be obeyed. The quarterly retraining audit assumed a change cadence the vendor didn't offer, a testing window the change process couldn't fit, and a cost line no one had funded. The risk rule was mathematically impossible given the operational reality underneath it — and nobody had noticed, because each pillar had been designed in its own room, by its own owner, against its own logic. The framework was complete. It was also inert.

That committee did not have a governance problem. It had a *systems* problem. And the difference between those two sentences is the whole of this chapter.

By the time you reach this point in building an enterprise AI capability, you've assembled the pieces. It is worth putting all fifteen on one page, because everything that follows depends on your being able to see them at once: **decision rights**, **accountability**, **board oversight**, **model risk**, **data quality**, **vendor strategy**, **architecture**, **monitoring**, **operational risk**, **cost control**, **capability**, **adoption**, **measurement**, **compliance**, and **change management**. Each is a distinct discipline, each earned its own hard-won chapter or three, and each is necessary. But if you've been paying attention, you've noticed that these pieces refuse to stay in their boxes. A decision about which vendor you lock yourself into constrains what you can afford to spend. Your cost model dictates which vendors you can even afford to evaluate. Your change-management process determines whether your governance rules are followed. Your data-quality framework decides how much risk your models actually carry. Everything touches everything else.

Here is the reframe this chapter is built on.

The myth: AI governance is a checklist of pillars, and you have governance when you've checked every box.

The reality: the fifteen pillars are one interconnected operating system, and failure comes from misalignment between pillars, not from missing pieces.

Most organizations are still running the checklist. Risk framework: check. Vendor approval process: check. Compliance audit trail: check. Change management: check. We have governance. It's a comforting inventory, and it is exactly the wrong mental model, because a checklist doesn't survive contact with reality. The moment something goes wrong (a model drifts, a vendor raises prices, a new regulation lands), you discover the gaps. They are never missing pieces; they are the places where the pieces were never wired together.

A Checklist Doesn't Survive Reality

Start with why the checklist model fails, because the failure is subtle and it feels like success right up until it isn't.

A checklist treats governance as a list of components you either have or don't. It has no way to represent the relationships *between* components, and the relationships are where governance actually lives. The insurer's impossible retraining rule wasn't a missing box; model risk, change management, vendor strategy, and cost control were all present and accounted for. The failure was in the seams. Four correct pillars produced one incoherent instruction, and no box on any checklist would ever have flagged it, because a checklist can't see a seam.

This is the recurring shape of governance failure at scale, and once you learn to see it you'll see it everywhere. It rarely looks like a hole. It looks like two reasonable rules colliding. Your fraud model needs to retrain fast to keep up with attackers, but your compliance process requires documentation that takes three weeks. Your architecture team wants a single shared platform for efficiency, but your risk team wants isolation so one failure can't cascade. Your product org is rewarded for shipping, but your accountability structure punishes whoever's name is on the incident. None of these is a missing pillar. Every one of them is a misalignment, and misalignment is what actually takes systems down.

So the question that matters is not "do we have all fifteen pillars?" You probably do, or you're close. The question is "do the fifteen pillars point in the same direction?" That is a systems question, and it has a systems answer.

Fifteen Pillars, Six Clusters, One System

The pillars don't connect randomly. They cluster into functional groups, and each group produces a capability the individual pillars can't produce alone.

Decision rights determine how fast you can move. But **cost control** constrains what you're allowed to decide, and **vendor strategy** determines which decisions are even on the menu. Together these three are your *choice architecture*: the set of moves actually available to you at any moment.

Model risk tells you what could go wrong. But **data quality** determines how much risk you're really carrying, and **architecture** determines whether you can change your risk profile later or you're stuck with it. Together these three are your *risk envelope*: the boundary of what you can safely hold.

Accountability tells you who answers when things break. But **board oversight** determines whether anyone above the incident set a risk appetite worth answering to, and **compliance** determines whether the answer you give is even enough to satisfy a regulator. Together these three are your *accountability surface*: the map of who is responsible for what, and whether that map is real.

Capability determines how much of the work you can do yourself instead of renting it. But **adoption** determines whether anything you deploy changes a single decision, and **measurement** determines whether you can tell the difference. Together these three are your *investment discipline*: whether the money you spend turns into anything.

Change management governs how you deploy updates. But **monitoring** determines whether you find out when those updates misbehave, and **operational risk**

determines what happens next when they do. Together these three are your *deployment reliability.*

Architecture determines whether you can isolate a model you no longer trust. But **vendor strategy** determines whether you're trapped with a choice you can't walk away from, and **model risk** tells you which of those choices you'd most regret. Together these three are your *strategic flexibility.*

Notice what happened in that last cluster: it introduced no new pillars at all. Strategic flexibility is built entirely out of three you had already placed somewhere else. That isn't sloppy taxonomy. It's the proof. A pillar that belongs to exactly one cluster is a component you could bolt on and forget. Architecture, vendor strategy, and model risk each anchor two clusters at once, which makes them load-bearing elements of a single structure. A change you make to one for a local reason ripples into a place you weren't thinking about. Harden your architecture for isolation and you have moved the walls of your risk envelope and bought yourself strategic flexibility in the same stroke, whether or not you meant to. Consolidate onto one vendor to simplify your choice architecture and you have spent the option to leave. None of these pillars works in isolation. They only work aligned.

What Alignment Actually Means

Alignment is an overused word, so be precise about what it means here, because the intuitive definition is wrong.

Alignment does not mean everything is one color, follows one process, or runs at one speed. It means the pressures point in the same direction. Here's what misalignment looks like in practice. Your decision-making process is designed to encourage speed. Your risk framework requires careful, deliberate analysis. Your change-management process mandates gradual, staged rollout. And your cost-control process penalizes the engineering overhead that careful testing demands. Four pillars, four sensible designs — and a question the organization re-litigates every single time, because the answer depends entirely on who is loudest in the room that day. Is it *move fast* or *be careful*? Nobody ever decided, so everybody keeps deciding.

Alignment means you decided it once. You said: "In this organization, at this stage, with these constraints, the right answer is X." And then you designed every pillar to point toward X.

If the right answer is *move fast*, your change-management process is automated, your compliance process pre-approves known-good patterns, your decision rights are delegated to the edge, and your cost model rewards quick iteration. If the right answer is *be careful*, your change-management process has human review gates, your compliance process requires documentation, your decision rights are centralized, and your cost model budgets for thoroughness rather than punishing it. Both of these are coherent. Both of them work. What fails is the third thing — the one most enterprises actually have, where half the pillars were tuned for speed and the other half for caution, by different people, at different times, none of whom compared notes.

This is why the systems view consists almost entirely of work that nobody schedules. Nobody gets promoted for reconciling the cost model with the change-management SLA. The single most useful hour I have spent inside a governance program was not a framework review; it was putting a platform-engineering lead and a model-risk lead in a room with their two documents open side by side and reading the clauses aloud until both of them could see that one required a rollback window the other's release process did not permit. Neither had written anything wrong. Neither had ever read the other's document. It took forty minutes to find the contradiction and one decision from the VP they shared to resolve it. And until that meeting happened, every deployment argument between those two teams had been a proxy fight over a conflict nobody could name. That is the shape of the work: two pillar owners in a room, forced to notice that their rules contradict each other, and then made to pick. The exciting version of AI governance (the frameworks, the principles, the maturity models) is the part your organization has already finished. The part that determines whether any of it functions is the quiet reconciliation between pieces that were built to be admired separately.

The Hidden Assumption Underneath Every Framework

There's something deeper going on when governance misaligns, and it's worth naming because it explains the resistance you feel from your own organization.

Almost every governance framework is built on an assumption that nobody states: *that all the pressures point toward compliance.* Design the controls, the thinking goes, and the organization will want to follow them. But that's not the world you operate in. Your organization also has a pressure toward speed, a pressure toward cost, a pressure toward innovation, a pressure toward taking risks that might pay off.

Governance does not eliminate those pressures. It cannot. Its actual job is to manage the tension between them.

So your real work is not to make the tension disappear. It's to make the tension *visible*, to state plainly what you're trading off, and to build systems that manage the trade-off the same way every time. That consistency is the whole product. And it's what separates governance that works from governance that breeds resistance.

Because here is the thing you already know if you've ever tried to roll out a governance process: people don't resist governance because it's hard. They resist it because it feels *arbitrary*. They resist because one day they're told to move fast and the next day they're blocked by a gate that demands caution. Because the compliance rule is airtight but the cost rule contradicts it. Because accountability points squarely at them while the decision authority sits two levels up. Resistance isn't a compliance failure or a culture problem. It's usually a correct read of a misaligned system. Your people can feel the contradiction even when they can't name it, and they're right to.

When your pillars are aligned, that resistance mostly evaporates, not because the rules got easier, but because they got *legible*. People will accept a constraint they can see the logic of, even when it costs them something. What they won't accept, and shouldn't, is a system that asks them to obey two rules that can't both be true.

The System Doesn't Hold Still

Everything so far assumes a fixed target: decide the right answer, align the pillars to it, done. But there's a second dimension to the systems view, and missing it is how organizations that *did* align their governance still end up buried. The target moves. Specifically, it moves as you scale. The governance that fits your first AI system is a poor fit for your tenth and an actively dangerous fit for your hundredth.

I've watched enterprises go from their first AI system to their tenth to their hundredth, and the governance challenge does not grow linearly with the count. It transforms. Most enterprises don't see this coming. They build governance for one system, it works beautifully, they deploy a second and discover the governance doesn't stretch, so they bolt on more process. By twenty systems they're drowning in overhead. By fifty they're paralyzed. The enterprises that navigate this well are the ones that understand what actually changes at each threshold, and rebuild governance on purpose instead of accreting it by reflex.

With one to five systems, governance is personal. You know the models. You know the data. You know the teams. You know what can go wrong, and you can still hold every conversation in one room. Governance is mostly four questions: Have we thought about the risks? Is there a human in the loop before this touches a person? Do we have monitoring? Can we turn it off? This is genuinely good governance. But it runs entirely on a small group of people who know everything about a handful of systems, and that is precisely what you're about to lose.

Between five and twenty systems, the room breaks. Now the systems live in different domains, on different data, carrying different risks, built by different teams. The patterns that worked before (constant synchronous communication, informal decisions made by the people who happened to be present) stop working the moment governance and model teams aren't in each other's pockets all day. So you need written standards, so that teams who never talk still know what's expected. You need decision gates with explicit criteria, so "approved for deployment" means something specific. You need escalation paths for the edge cases. And you need governance infrastructure that doesn't collapse when one key person is on vacation. You've stopped doing governance *thinking* and started building governance *processes.*

Between twenty and fifty systems, one-size-fits-all dies. Now you have specialized domains (lending systems, pricing systems, recommendation systems), each with its own risk profile, data sensitivity, and regulatory weight. A single uniform process can't fit all of them without being either too heavy for the light cases or too light for the heavy ones. What works instead is principles-based governance: set the principle, let teams implement it differently. "Any system that makes decisions affecting customers must have human review" is a principle. But *human review* means different things in different places. I watched a payments company write that sentence once and implement it three ways inside a single quarter. In the lending model, it meant the model recommends and a human decides. In the recommendation engine, it meant a team manually auditing the top decisions each week. In the pricing service, it meant a rule that automatically blocks outlier decisions before they ship. Same principle, three implementations, and all three were defensible to the same regulator. This demands frameworks that are principles-based rather than process-based, explicit mechanisms for tailoring the framework to each domain, and someone empowered to make the risk-versus-speed-versus-cost trade-off for that domain rather than pretending one balance fits all.

Past fifty systems, you're running a portfolio. You're no longer governing individual systems; you're managing a portfolio of them, built at different times, on different frameworks, by teams with wildly different governance maturity. Some drive revenue. Some reduce risk. Some are experiments that may not survive the quarter. Your governance has to reflect those different roles rather than treating every system as equally precious. And you cannot rebuild governance from scratch for each new system, because there are a hundred of them and they don't share a baseline. This is where you need portfolio-level standards that span different maturity levels, *progressive* governance that can raise the bar on existing systems without forcing a full re-architecture, ruthless simplification (a hundred systems cannot mean a hundred bespoke governance approaches), and incentive structures that make doing governance right easier than skipping it. Because at this scale, if the right path is the hard path, people will route around it.

What Actually Works

Across all of that, the enterprises that scale governance well do three things, and they're the same three things whether you're crossing from five systems to twenty or from fifty to a hundred.

First, they anticipate the evolution. When they have three systems, they're already designing for twenty: building infrastructure that can grow, choosing tools that scale, making architecture decisions with future volume in mind. Second, they rebuild intentionally at the inflection points instead of bolting on more process. Crossing from a handful of systems to a few dozen, they don't add another approval stage; they redesign the whole approach. Different structure, different tooling, different organizational shape. Third, they invest in governance infrastructure before it's obviously necessary. Governance platforms, monitoring tooling, decision-support systems: these look like pure overhead when you have one system and turn out to be oxygen when you have fifty.

And they avoid the most common mistake, which is seductive precisely because it feels responsible: scaling governance by replicating the same governance for every system. This works right up until scale makes it impossible, and then it fails in one of three ways. You build crushing overhead, where every system grinds through the same ten-week approval regardless of whether it's a fraud model or a meeting-notes summarizer. Or teams quietly abandon governance because it's too slow to be worth it. Or you get governance anarchy, where every team invents its own approach and nothing is comparable. None of the three works. What works is governance that evolves with scale on purpose, not by accident.

So locate yourself honestly on that same ladder. With one to five systems, build governance *thinking* that can survive scale: invest in written standards and clarity now, even though informal communication would still get you through today, because informal communication is exactly what breaks first. With five to twenty, you're standing on an inflection point: look hard at whether your governance is scaling or creaking, and if it's creaking, rebuild it before it fails rather than after. With twenty to fifty, you need portfolio thinking: one set of standards that has to hold across systems built in different years by teams at different levels of maturity, which means writing the standard as a principle and letting each domain implement it. Past fifty, you need progressive governance, because you can no longer force every system into one framework and the attempt will only produce the overhead, the abandonment, or the anarchy.

The two dimensions of the systems view are really one instruction. Align your pillars so they point the same direction, and then accept that the direction, and the machinery that enforces it, has to be rebuilt at each threshold as you grow. The complexity you're managing was never new. You've been managing it all along. You just weren't thinking of it as one system. Once you do, everything gets clearer — and a great deal harder to ignore.

Knowing the pillars form a system is not the same as running one. Chapter 23 is the operating rhythm: what to audit before go-live, and what to check ninety days after.

FIGURE 22.1

The Fifteen Pillars Operating System

FIFTEEN PILLARS	WHAT IT DETERMINES	SIX CLUSTERS
01 **Decision rights**	how fast you can move	**Choice architecture** the moves actually available to you
02 **Cost control**	what you're allowed to decide	
03 **Vendor strategy**	what is even on the menu	CLUSTER SIX · LOAD-BEARING **Strategic flexibility** Adds no new pillars — these three each anchor two clusters at once.
04 **Model risk**	what could go wrong	
05 **Architecture**	whether you can change later	
06 **Data quality**	how much risk you carry	**Risk envelope** the boundary of what you can hold
07 **Accountability**	who answers when it breaks	**Accountability surface** who answers for what, and if it's real
08 **Board oversight**	whether an appetite is set	
09 **Compliance**	whether a regulator accepts it	
10 **Capability**	how much you do yourself	**Investment discipline** whether the money turns into anything
11 **Adoption**	whether anything changes	
12 **Measurement**	whether you can tell	
13 **Change management**	how updates ship	**Deployment reliability** whether updates ship without damage
14 **Monitoring**	whether you find out	
15 **Operational risk**	what happens next	

Tune one pillar and the others move.

Vendor strategy, model risk and architecture each anchor two clusters — trace the ripple first.

Workshop use: place a sticky per real pillar, then draw a line wherever one serves two clusters.
Anything with two lines is load-bearing — trace where the ripple goes before you tune it.

FIGURE 22.1

FIGURE 22.2

The Governance Scaling Ladder

Each rung is 10× the last — and each needs a different mechanism, not more people.

1 system — **Personal and synchronous**

GOVERNANCE LOOKS LIKE	WHAT BREAKS	WHAT MUST CHANGE
One owner, one room. The decision gets made in conversation, not on paper.	Nothing — which is the trap. Working without process reads as proof you never did.	Write the first decision record: one page naming who decided what, and why.

PHASE CHANGE · JUDGMENT BECOMES STANDARD

10 systems — **Written decision architecture**

GOVERNANCE LOOKS LIKE	WHAT BREAKS	WHAT MUST CHANGE
A standing review board, risk tiers, and a shared register of decisions and owners.	Memory and calendar. No one holds ten sets of thresholds; review turns into a queue.	Codify tiers, owners and thresholds until the governed path is the easiest path.

PHASE CHANGE · STANDARD BECOMES CODE

100 systems — **Automated and instrumented**

GOVERNANCE LOOKS LIKE	WHAT BREAKS	WHAT MUST CHANGE
Policy runs in the pipeline. People handle exceptions, not approvals.	Review itself. A human queue is a bottleneck no amount of headcount can clear.	Instrument the estate: govern by exception and portfolio signal, not case by case.

HEADCOUNT CANNOT CROSS A PHASE CHANGE

Workshop use: pin your live system count on a rung, then read the rung above it.
The gap between those two rungs is your build list — mechanisms to ship, not roles to hire.

FIGURE 22.2

The Bottom Line

The fifteen governance pillars are not a checklist; they are one interconnected operating system, and enterprise governance fails at the seams between pillars far more often than it fails from a missing piece. A framework where every box is checked can still issue an instruction no one can follow, because four correct pillars can combine into one incoherent rule, and no checklist can see a seam. Alignment is the fix, but alignment doesn't mean uniformity; it means deciding once whether your organization's answer is *move fast* or *be careful* and then tuning every pillar to point that way, because both coherent answers work and only the misaligned middle fails. The work of getting there is quiet reconciliation between pieces that were built to be admired separately, which is why most organizations skip it and why the ones that do it pull ahead. And because the target moves as you scale from one system to a hundred, the operating system has to be rebuilt intentionally at each threshold; the governance that fits your tenth system will bury your hundredth, and the enterprises that win are the ones that rebuild before they're forced to.

Field Kit

1. The seam-check question set. Bring these to your next governance or architecture review and force real answers, because every one of them probes a place where correct pillars can contradict each other: - Is there any rule in our framework that our operational reality — vendor cadence, testing windows, budget lines — makes impossible to actually follow? - Where do two of our pillars issue instructions that can't both be obeyed at once (e.g., "retrain fast" vs. "document for three weeks")? - If we asked ten people whether our organization's answer is "move fast" or "be careful," would we get one answer or ten? - Which pillars were tuned for speed, and which for caution — and did anyone ever compare the two lists? - When people resist a control, are we treating it as a culture problem, or checking whether they're feeling a real contradiction?

2. The alignment declaration (one line, then a checklist). Write the single sentence out loud: *"In this organization, at this stage, with these constraints, the right answer is ___."* Then verify each pillar points there. For "move fast": is change management automated, are compliance patterns pre-approved, are decision rights delegated, does the cost model reward iteration? For "be careful": does change management have review gates, does compliance require documentation, are decision rights centralized, does the cost model fund thoroughness? Any pillar pointing the other way is your misalignment, and it's costing you more than you think.

3. The six-cluster map. Take the fifteen pillars — decision rights, accountability, board oversight, model risk, data quality, vendor strategy, architecture, monitoring, operational risk, cost control, capability, adoption, measurement, compliance, change management — and place each one into the clusters it belongs to: *choice architecture* (decision rights, cost control, vendor strategy), *risk envelope* (model risk, data quality, architecture), *accountability surface* (accountability, board oversight, compliance), *investment discipline* (capability, adoption, measurement), *deployment reliability* (change management, monitoring, operational risk), and *strategic flexibility* (architecture, vendor strategy, model risk). Mark the three that appear twice — architecture, vendor strategy, model risk. Those are load-bearing: a change you make to one for a local reason ripples into another cluster. Before you tune any of them, trace where the ripple goes. Then argue about the placements; the argument is the exercise.

4. The scale-stage locator. Place your portfolio on the ladder and name the next threshold: **1–5 systems** — governance is personal; build standards and clarity now even though informal comms still work; **5–20** — the room breaks, so decide whether governance is scaling or creaking and rebuild if it's creaking; **20–50** — one-size-fits-all dies, so move to portfolio-level standards written as principles that each domain implements its own way; **50+** — adopt progressive governance that raises the bar on existing systems without re-architecture, because forcing one framework onto everything will produce overhead, abandonment, or anarchy. Write down which transition you're closest to and what you'll rebuild *before* it forces your hand.

5. The principle-to-implementation template. For each governance principle that must scale across domains, fill three columns: *Principle* (the fixed rule, e.g., "customer-affecting decisions require human review"), *Domain* (lending / pricing / recommendation / …), and *Implementation* (what the principle concretely means there — "model recommends, human decides" vs. "block outliers by rule" vs. "weekly manual audit"). Same principle, different rows. If a domain has no row, it's either ungoverned or governed by an accident.

6. The re-architecture trap diagnostic. Ask which of the three failure modes you're drifting toward as you scale: crushing overhead (every system through the same heavyweight process), quiet abandonment (teams skipping governance because it's too slow), or anarchy (every team inventing its own approach). Whichever you can feel starting is the signal that you've hit an inflection point and are bolting on process instead of rebuilding. Name it before it hardens.

CHAPTER 23

Putting It Into Practice: The Readiness Audit & the 90-Day Journey

Part VIII · From Pieces to System

When a payment processing system fails, there is a protocol. When a trading system misbehaves, alarms go off and someone whose actual job is monitoring picks up the phone. When supply chain optimization goes wrong, there is a decision structure for who investigates, who decides, and who is accountable to whom. These are solved problems. Every critical business domain has an answer to *what happens when this breaks*, and the answer is boring, rehearsed, and owned.

When an AI system breaks, many organizations are improvising.

Not because they lack commitment to doing the right thing. Because they built the technical capability far faster than the organizational capability to manage it. You will find a company with leading-edge machine learning, models that beat every internal benchmark, beautiful infrastructure — and a decision-making process that would be unrecognizable in any other high-stakes part of the same business. Sophisticated systems, running inside primitive governance structures.

That gap is the central problem in enterprise AI. Not the models. The decision-making. The models work, the integration is clean, the technical problems are getting solved. What is not getting solved is the organizational ability to *see* what you have built, make deliberate decisions about how to use it, and respond when those decisions need to change.

That is the whole chapter, really. Everything that follows argues that the shape of this problem is different than most organizations assume, and that two practices, a readiness audit before go-live and a ninety-day checkpoint after, are how you turn that insight into an operating rhythm instead of a slogan. Both are specified here as things you can put on a calendar, because a practice that is not on a calendar is a preference.

Here is the reframe this chapter is built on.

The myth: once the models work and the pre-deployment compliance audit passes, the organization is ready to run the system in production.

The reality: the gap is organizational, not technical, and you close it with two dated practices: an audit that interrogates operations before go-live and a ninety-day checkpoint after.

Technical Readiness Is Not Governance Readiness

Start by separating two things that constantly get confused. You can have sophisticated models, a data science team that knows how to train and ship them, and good technical monitoring: dashboards showing accuracy, latency, throughput. None of that is governance.

Governance answers a different set of questions. Who decided this model was appropriate for this use case? What assumptions does that decision rest on? If those assumptions change, who notices, and who decides what happens next? If the model fails, who is accountable, to whom, and for what? Most organizations have thorough answers to the technical questions and answers to maybe thirty percent of the governance ones.

The gap shows up in a handful of predictable ways, and once you see them you cannot unsee them.

Responsibility does not distribute automatically. Organizations assume that because a system is deployed and running, someone owns it. In practice ownership is ambiguous: the business leader thinks the technical team has it, the technical team thinks the business leader has it, and the governance team is advisory with no binding authority over anything. When something finally needs to happen (an investigation, a remediation, a decision to change course), nobody is sure who is allowed to start it, and weeks evaporate while the organization reverse-engineers its own authority structure. The organizations without this problem name three seats per system in writing: who owns the decision, who owns the data, and who can pause it. You know from Chapter 1 what shape each answer takes. The third seat is the one that stays empty, because naming someone who can stop a system the business wants running admits that stopping might one day be correct. So it stays implicit, which means it belongs to whoever escalates hardest, which means nobody. A pause seat that functions has a written trigger, no requirement to convene anyone before acting on it, and a holder who has used it at least once. Unexercised, it is not authority; it is an assumption about authority.

Observability is not governance, though most organizations treat them as the same purchase. Chapter 12 made the case that watching a model's behavior is its own discipline. The readiness version is narrower: a dashboard is a purchase and a decision structure is an org design, and only one of them has a vendor who will sell it to you. So the drift alert fires exactly as designed, lands in a channel shared by four teams, three of whom assume the fourth owns it, and scrolls away inside a day. The signal was perfect; the chain from signal to decision to action never existed. Before approving the monitoring spend, get names rather than teams against three questions: who reads this alert, who decides whether it matters, who is permitted to act.

Nothing on the calendar forces a stop. Chapter 15 supplied the question: if we were starting today, knowing what we now know, would we do this? The reason organizations never ask it is structural rather than psychological. The original sponsor moves to another division. The business case was written against a vendor price that has since fallen by two-thirds and a capability gap the base models closed on their own. Ask three people on the program privately and none will argue it is still the best use of the money — and not one has an occasion at which saying so is their job rather than sabotage. Absent a scheduled moment, the default is continue, and continue never has to be defended.

You cannot measure what you refuse to define. Chapter 17 settled what "adoption" has to mean: decision change, not usage. The readiness point is about timing. The decision-change test needs a baseline, and a baseline is the one input you cannot retrofit: once the system is live in the workflow the counterfactual is gone, and every measurement after that is an argument rather than a number. So the definition belongs in the pre-deployment audit alongside monitoring and fallback: what decision is this meant to change, what do those decisions look like now, and who is holding out a group or a stretch of history to compare against? Teams push back because it delays launch by a few weeks. The alternative is a whole system lifetime spent unable to say whether it did anything.

Notice what all four have in common. Each is a missing link between information and action: the signal exists, or easily could, and no named person, scheduled moment, or prepared comparison converts it into a decision. Governance is that link:

continuous operational visibility attached to somebody who can act on it. It is unglamorous, and it is the actual work.

The Audit You Think You Need, and the One You Actually Need

Here is where the abstraction becomes a meeting on your calendar. Suppose you are planning a significant deployment: a new recommendation system, a content moderation tool, an internal automation that touches real decisions. Your governance team is probably already planning an audit. Good. The uncomfortable question is whether that audit will actually tell you if you are ready.

Most enterprise AI audits are built to check the boxes that are easy to check. A typical pre-deployment audit asks whether the model meets performance thresholds, whether the vendor is SOC 2 certified, whether the data inputs have been classified, whether there is explainability documentation, whether a risk rating exists. These matter. But notice that nearly all of them are verifiable *without touching the system.* A vendor can hand you a security audit; a classification template can be filled in; a risk matrix can be drawn in a slide deck. You can pass every one and still have no idea whether you are ready to run this thing in production, because readiness is not primarily about documentation. It is about visibility.

The audit that predicts failure asks fundamentally different questions. Not *have we documented the risks?* but *do we actually understand what happens when this system fails?*

- What does failure look like for *this specific system?* Not "model performance degrades," but: what does the business see? How long before anyone notices? Who notices first: a customer, a regulator, or you?
- What is the path from *something is wrong* to *we know something is wrong?* Do you have monitoring in place, and what signals are you actually tracking?
- If you turn this off today, what happens? Is there a fallback? How long does it take to activate? Does anyone even know it exists?
- What happens to the data? Not "we comply with GDPR," but: where does input data sit, for how long, who can access it, and what happens if a deletion request arrives at three in the morning while the system is mid-inference?

- What won't this system tell you? What are the blind spots you will only discover after you have already had problems?

These questions are hard precisely because they require thinking through operations rather than compliance, and there are structural reasons most audits duck them. The audit happens before deployment, when there is no production data and no real failure to analyze, so it defaults to *do you have a plan to monitor this?* instead of *show me the monitoring.* Most enterprises also lack the right people in the room; governance and compliance leaders are fluent in data classification and vendor security and far less comfortable interrogating observability architecture or failure modes, so the audit drifts toward what the audit team knows how to verify. And there is an asymmetry of effort: a vendor produces a security checklist in an afternoon, while building production monitoring for a system that does not exist yet is genuinely hard. So audits reward thoroughness on the easy half and speed-run the half that matters.

When you push on the harder questions, you frequently discover you are not ready — not in a compliance sense, because most large enterprises can pass a compliance audit in their sleep, but in an operational one. The system has no monitoring. There is no agreed fallback. Nobody has written down what "failure" means for this business. Data handling during errors is a shrug. These are almost always fixable, but only cheaply if you find them *before* deployment. Finding them six months in, when the system is embedded in a dozen workflows, is a far more expensive project.

So do the audit differently. Start with operations, not compliance. Ask the team that will *run* the system (not the team that will govern it) what they would need to know to be confident, and what would tell them something is wrong. Make the answers specific. Not "monitor model performance" but "track prediction confidence by input type weekly, and alert if any segment drops below X." Not "ensure data security" but "describe the 72-hour retention policy after inference completes, the access controls on inference logs, and exactly what happens when a CCPA deletion request lands." Have them build the monitoring *before* you audit, so you are evaluating actual monitoring and not a document about monitoring. The difference is decisive. Ask them to describe failure in scenarios (a customer problem, a compliance problem, an accuracy problem), and for each, the path to detection and the response. Produce the fallback runbook Chapter 13 specifies, in numbers rather than intentions, with one named owner and a date on the last drill. And be honest about the unknowns: surface them explicitly rather than pretend you have thought of everything.

The earliest problems with AI systems are usually not problems with the models at all. They are operational. You did not realize how long inference takes, so the system introduces latency nobody budgeted for. You did not account for the volume of edge cases, so the fallback is overwhelmed the first time it is needed. You did not monitor the right signal, so by the time you notice something is wrong it has been wrong for months. An operational audit does not prevent every one of these; it changes which problems you catch early and which ambush you in production. Most enterprises will pass a traditional audit before deploying. The question is whether they will also pass an operational one.

The Ninety-Day Checkpoint

The audit gets you to go-live and says nothing about what happens after, which is where governance either becomes an operating rhythm or evaporates. So pair it with a second practice, deliberately unimpressive: ninety days after a system goes live, and every ninety days after that, the people who own it sit in a room for ninety minutes and answer the same four questions.

Ninety days is not a magic interval; it is roughly the shortest one at which the answers change. Under a month you are still reading launch noise: novelty usage, hypervigilant operators, a sponsor watching. At six months, drift has had two quarters to compound and the sponsor has moved on. Ninety days is about one business cycle: long enough for real usage patterns and one surprise, short enough that nothing large has been built on an assumption that stopped being true.

Who is in the room is the part organizations get wrong. The three named seats (decision owner, data owner, pause-authority holder) plus one or two people who operate the system daily and can say what it has been doing, and optionally one governance person there to interpret risk rather than approve anything. Four to six people; let it grow past that and it becomes a steering committee, which is the failure mode.

The agenda is fixed, which is the only reason it survives a busy quarter:

- **Do the assumptions still hold?** Read the register and mark it up. Assumption coherence is the measure, checked against what the last ninety days looked like rather than against the deployment memo.

- **What did the system do that surprised us?** Not incidents. You know about incidents. Surprises are what did not trip anything: the segment nobody expected to dominate traffic, the workaround ops invented in week three and never mentioned.
- **Did it change any decisions?** The decision-change test, run against the baseline you were forced to capture before launch.
- **Would we start this today?** Asked last, with the first three answers on the table.

The artifact matters more than the meeting. One page: three findings, each with a decision, a date, and a name (continue as-is, change this specific thing by this date, or stop). "Keep monitoring" is not a decision. The previous page gets read first, and the first item of business is what happened to it. Nobody builds slides; the moment someone does, it has become a place to present rather than decide, and the operators stop saying inconvenient things.

Run this across a portfolio for a year and it produces what no framework does: a written record of how your organization's thinking changed. That is the real output. It is where you find out that the risk the program was designed around was not the risk you ran, and that a control everyone believed in has never once fired.

Across enough of these, three things move. The reframe Chapter 19 names, from *what are we required to do* to *what are we assuming*, becomes the first agenda item rather than executive language. Governance migrates from after the fact to before commitment, and decision velocity (Chapter 3) is the honest read on whether that worked: slow means you built a bottleneck, fast means you built a capability. And the questions start pointing forward, which is Chapter 14's reactive-to-predictive shift arriving on its own rather than being installed.

The Tensions That Never Resolve

Run enough checkpoints and you notice a second thing: some findings keep reappearing on the page and never close. Those are not failures of the practice. They are permanent tensions: features, not bugs. The only real question is whether you are managing them consciously or letting them manage you.

Speed versus depth. You cannot move fast and understand everything deeply; you choose where to go deep and where to move fast. Strong organizations say it out loud: *this system needs deep analysis, this one does not, and here is why.* Weak ones pretend the choice does not exist and do both poorly.

Centralization versus distribution. Keep all governance expertise in one place and you cannot be fast. Scatter it everywhere and you cannot stay consistent. The right balance is specific to your organization and it moves as you mature. The signal of health is not sitting at some perfect point. It is knowing where the point currently sits.

Compliance versus competitiveness. Some of what you do for compliance creates no competitive advantage, and some of what creates advantage runs ahead of what compliance requires. Leaders manage both explicitly instead of pretending they are the same thing.

Control versus trust. The more you control through process, the less you rely on teams to decide well on their own. The more you trust teams to decide, the more you need confidence they are deciding with full information about risk. It is a trade-off you either manage consciously or absorb blindly.

What This Means for You Right Now

Pull it together and four instructions fall out.

Build governance that fits your risk profile, not your risk theater. The size of your program should match the risk you are actually creating, not the anxiety you feel about AI. A company running one customer-facing model in a regulated industry may need a more sophisticated structure than a company running fifty internal tools.

Give both practices a date and a name. An audit nobody schedules is a philosophy, and a checkpoint nobody owns gets skipped the first quarter it collides with a launch — and then the one after, because nothing bad visibly happened the first time. Put the audit on the deployment plan as a gate rather than a milestone, book the first checkpoint on the day you go live, and staff both with the distributed capability Chapter 16 describes doing its actual job. Keeping them on the calendar through a busy quarter is the hard part, and the only part that compounds.

Measure governance differently. Stop counting completed compliance checklists. Track decision velocity, assumption coherence, and the ratio that matters most:

problems caught early against problems discovered late. That last one is the only number on the list that improves purely because the checkpoint exists, which makes it your cleanest evidence the practice works.

Be explicit about your trade-offs. Where have you chosen speed over depth, and where does depth win? Where are you managing the tension between compliance and competitive advantage? Name these choices and communicate them. The worst governance programs are the ones where the trade-offs are implicit. People argue endlessly because no one ever admitted a choice was being made.

Underneath all of it is a single reframe. Enterprise AI governance is not primarily a technical problem or a compliance problem. It is an organizational learning problem. The organizations pulling ahead are not ahead because their policies are better; they are ahead because people understand the assumptions they are making, risk is visible before decisions lock in, and they can move fast precisely because they understand their own risks. That is not infrastructure you implement once. It is a thinking problem, and thinking problems require continuous change rather than a launch date.

The transition every early-stage effort is trying to make is from *we have capability* to *we have managed capability*, and it is never automatic. It requires deliberate design, and something harder to install: cultural permission to ask the uncomfortable questions. To say *should we still be doing this?* To admit an assumption was wrong. To tolerate stopping a program that is not working. That is harder than building the models. The systems are ready. The real question, the one the audit and the checkpoint exist to answer, is whether the organization is.

The Bottom Line

The gap in enterprise AI is not a capability gap; it is a governance gap: the organizational ability to see what you have built, decide deliberately how to use it, and respond when the decision needs to change. You close it with two practices, both of which fit on a calendar. The readiness audit runs before go-live and interrogates operations rather than documentation: what failure looks like here, the path from wrong to known-wrong, the fallback and when it was last drilled, where the data sits, and the baseline you will need later to prove the thing changed anything. The ninety-day checkpoint runs after and keeps running: ninety minutes, the three named seats plus the people who operate the system, four fixed questions, one page of findings carrying a decision, a date, and a name. The audit tells you whether you are operationally ready to deploy; the checkpoint tells you whether your organization is learning, because a year of those pages is a record of how its thinking changed. The systems are ready — the work is making the organization ready to run them.

Field Kit

1. The operational-readiness audit (five questions for any pre-deployment review). Ask the team that will *run* the system, not the team that will govern it: (a) What does failure look like here — what does the business see, how long before anyone notices, who notices first? (b) What is the path from *something is wrong* to *we know something is wrong*? (c) If we turn this off today, what is the fallback, how long to activate, and does anyone know it exists? (d) Where does input data sit, for how long, who can access it, and what happens to a deletion request at 3 a.m. mid-inference? (e) What won't this system tell us?

2. The "show me, don't tell me" rule. Before the audit, require the running team to have built the actual monitoring and stated it in numbers — not "monitor model performance" but "track confidence by input type weekly, alert if any segment drops below X." Same standard for the fallback: you are auditing Chapter 13's runbook, not a document promising one, and the only evidence that counts is the date it was last drilled.

3. The pause seat. Decision owner and data owner get named without much argument. The third seat — who can stop this system without convening anyone — is the one that stays empty. Fill it in writing, give it a trigger threshold, and confirm the holder has exercised it at least once. Authority that has never been used is authority you are assuming you have.

4. The ninety-day checkpoint. Ninety days after go-live and every ninety days after: ninety minutes; four to six people (the three named seats plus one or two daily operators); four fixed questions — do the assumptions still hold, what surprised us, did it change any decisions, would we start this today. Output is one page: three findings, each with a decision, a date, and a name. Read the previous page first and open with what happened to it. No slides. Three identical pages running means it has become theater.

5. The sunk-cost circuit breaker. Chapter 15's start-over question, on the calendar as question four above, so the default stops being "keep going" and someone has to defend continuing.

6. The adoption metric that isn't usage. Apply the decision-change test from Chapter 17; usage without decision impact is theater. The piece that belongs in the pre-deployment audit is the baseline — the only input you cannot capture after launch.

7. The four-tension dashboard. Once a quarter, force a conscious position on each permanent tension — speed vs. depth, centralization vs. distribution, compliance vs. competitiveness, control vs. trust — alongside decision velocity (Chapter 3) and assumption coherence (Chapter 7). You are not solving these; you are proving you manage them on purpose rather than by accident.

CHAPTER 24

Beyond Compliance: Governance as Advantage & What Comes Next

Part VIII · From Pieces to System

Somebody senior always asks the same question, and always in the tone of reasonable prudence: *"What's the minimum we need to do to comply?"*

It sounds responsible. It is, in fact, one of the most expensive questions you can ask, because it assumes that compliance is the constraint. It isn't. The constraint is everything compliance is trying to prevent: deploying systems that fail, that create liability, that damage trust, that lock you into operational patterns you can't escape. Compliance is a low-resolution shadow of that real problem. Aim at the shadow and you hit exactly what you aimed at: a program that passes audit and governs nothing.

The organizations getting this right have stopped thinking about governance as a cost and started thinking about it as operational infrastructure. Supply chain management isn't a cost; it's what lets you scale. Financial controls aren't a constraint; they're what make complex transactions possible in the first place. Governance becomes a competitive advantage at the exact moment you stop treating it as something imposed from outside and start treating it as something that lets your business move faster.

This is the closing argument of everything that came before, and it has two halves. The first is about the present: why the organizations that win treat governance as capability rather than cost. The second is about the future: what enterprise AI governance is about to become, and what you should be building now so that the program that is state-of-the-art today isn't obsolete in eighteen months.

Here is the reframe this chapter is built on.

The myth: governance is a cost to be minimized, so the right question is what the minimum is to pass audit.

The reality: aiming at compliance gets you a program that passes audit and governs nothing, while governance built as operational infrastructure is what lets your business move faster.

The Minimum-Compliance Trap

Companies that optimize for minimum compliance nearly all converge on the same architecture, and you can recognize it from across the building. There's a compliance team checking boxes. There's a business team moving fast and discovering the problems later. And there's a permanent low-grade friction between them.

The compliance team is cast as the people who slow things down. The business team is cast as the people who take reckless risks. They are, structurally, adversaries. And in that adversarial arrangement the whole organization moves *slower*, not faster. Then something goes wrong. The compliance team says "we told you so" and implements stricter controls, which makes everything slower still, which makes the business team route around them more aggressively, which produces the next incident. The cycle has its own momentum.

"What's the minimum we need to comply?" leads directly here, because the honest answer is almost always "enough to pass audit." And passing audit is a remarkably low bar for actual governance. You can pass an audit for a system you don't understand, can't monitor, and couldn't shut down cleanly if it started making bad decisions on Monday morning. The box gets checked. The risk stays exactly where it was.

What Competitive Governance Actually Looks Like

The organizations genuinely leading on enterprise AI have reframed the whole thing. Compliance is a constraint they manage. Table stakes, nothing more. Governance is for something else entirely: enabling confident deployment at scale. It's the infrastructure that lets you move fast *because* you understand your risks rather than in spite of not understanding them.

In practice that looks different from most governance programs. There is no elegant policy document at the center of it. There are four capabilities, and not one of them photographs well.

The first is **distributed decision-making**. Governance isn't an event that happens at a gate; it's a skill the teams doing the work have already built (Chapter 16's argument that capability is grown rather than hired). A gate is a bottleneck by design. Distributed judgment isn't.

The second is **predictive rather than reactive risk management**: Chapter 14's shift, running as the default operating mode rather than sitting on a maturity model as an aspiration, and faster than the alternative precisely because you aren't losing weeks to investigation and months to remediation.

The third is **decision frameworks that empower instead of constrain**. The weak version is "you need approval to deploy." The strong version is "if your system meets these criteria, deployment is automatic; if it doesn't, here's exactly why and what to change." Teams stop waiting for exceptions and start building systems that clear the bar on the first try. The framework becomes a target to hit, not a wall to negotiate.

The fourth is **learning that feeds back into architecture**. When something fails, you don't just investigate and close the ticket. You let the failure change how you build the next system like it. Over enough cycles your architectures improve, your decisions improve, and you simply fail less often. This is the compounding part, and it's invisible on any single quarter's dashboard.

Notice what all four have in common: none of them is exciting. The organizations moving fastest aren't the ones with the most sophisticated ethics framework. They're the ones that did the infrastructure work.

Speed and Safety Are the Same Goal

Here is where the framing flips from cost to capability. The myth is that you move fast *despite* governance: that safety is a tax on velocity, and the best you can hope for is to keep the tax low. The reality, in the organizations that have actually done it, is that they move fast *because of* governance.

They move fast because they understand their risks, so they can make confident decisions instead of hedged ones. They move fast because they catch problems early, so they don't surrender months to remediation. They move fast because they learn from what happens, so next quarter's decisions are better than this quarter's. They move fast because decision-making is distributed, so nothing sits in a queue waiting for a central committee. Every one of those is a governance capability. Every one is also, directly, a competitive advantage.

Compare that to the company that treats governance as a constraint. It moves fast at first: deploying systems it doesn't fully understand, booking the early wins. Then it discovers the problems. Then it has the incident. Then it explains the incident to a regulator. Then it bolts on strict controls that slow everything to a crawl. That is not speed. That is speed followed by a hard stop, and the hard stop erases the lead.

None of this is free. Building governance that's genuinely competitive costs more upfront than checking boxes: you need people with real governance expertise distributed through the organization, systems that give you visibility into your risks, and an investment in learning infrastructure that won't show a return next quarter. What you get back, over a multi-year horizon, dwarfs it: shorter time from decision to deployment, a falling incident rate, compounding organizational learning, regulators who extend lighter-touch oversight because they can see you understand your own risks, and customers who trust you because you've earned it. The gap between genuine governance and governance theater doesn't look like much in month one. Over three years it's enormous.

The Threshold Moment

There is almost always a specific moment when an organization crosses from "governance is a cost" to "governance is a capability." It is rarely triggered by an article or a consultant. It's triggered by a meaningful incident that cost something real: money, a marquee customer, regulatory attention, reputation.

At that moment the organization is forced to choose. One path is to double down on governance theater: hire more auditors, write more policies, add more gates. The other is to invest in real infrastructure: distributed capability, monitoring, better people. The ones that choose the second path tend to find, roughly six months later, that their incident rate has dropped and their decision velocity has *improved*: that they are moving faster and being safer at the same time, which the first path told them was impossible.

If you're reading this, you're probably past the minimum-compliance-trap version of your organization. You've accepted that governance matters. The open question is whether you still think of it as a cost to minimize or a capability to build, and there's a clean diagnostic for that. *Does governance slow you down or speed you up?* If every governance interaction adds time, you have theater. If governance interactions raise your confidence in decisions and cut the time you spend cleaning up after deployment, you have something real. The shift you're looking for isn't "more governance." It's governance that actually works. That, and not the policy binder, is the competitive advantage.

What Comes Next

Everything to this point has been about the present moment: the structural, procedural challenges of building frameworks that fit foundation models, distributing expertise, moving from reactive to predictive. Those are hard problems, but they're addressable with work, expertise, and organizational will. The challenges coming are a different shape.

What follows is not a set of predictions about AI itself. It's a set of observations about emerging patterns: scenarios to prepare for rather than certainties to bet on. But the direction is clear enough that failing to prepare is its own decision. It points to a world where AI isn't an isolated function but embedded everywhere; where the competition isn't about who has the best AI but about who can govern it reliably enough to earn trust; and where the risk surface isn't a model or a dataset in isolation but the entangled system of humans, algorithms, and business processes around it.

From Standalone Systems to Networks

Today most enterprises still treat AI systems as discrete things. One team owns the recommendation model, another fraud detection, another the customer-service assistant. Each governed on its own, with its own risk framework and its own line of accountability. That separation was always somewhat artificial, but it was manageable.

That is ending, for two reasons. First, enterprises are stacking systems: you call the recommendation model, which calls another model, which reads data a third system already transformed. The risk surface is no longer a model; it's a network. Second, AI is becoming general infrastructure rather than specialized application: the same foundational system used across the organization in dozens of ways, most of which no one centrally architected.

Picture the ordinary version. A pricing model consumes a demand forecast, which consumes embeddings from a shared internal service. The embedding service gets upgraded: a routine improvement, approved by the team that owns it and correctly reviewed against its own risk framework. Nothing downstream registers a change, because the pricing model's inputs still arrive well-formed and in range. Three months later margins are off by a point in one region and nobody can say why. Every system passed its own review. The network drifted anyway, and the drift lived in a seam on no team's risk register.

The second shape is reuse. A model gets approved for a low-stakes internal purpose (summarizing support tickets, say) with a review proportionate to that stake. Six months later the same endpoint sits in a customer-facing decision path, because it was available and it worked. Nobody re-reviewed it; from the platform's view nothing changed. The approval attached to the system. The risk attached to the use. Those two came apart, and most governance frameworks have no mechanism that notices.

So the question becomes one that barely exists today: how do you govern a network whose components are interdependent, where one model serves a dozen contexts, and where a failure in one place cascades? Frameworks built for individual systems will not scale to it. You need network-level governance: asking not just whether each system is safe but whether the *network* is stable, and tracing where a change propagates before you approve it. The practical starting point is smaller than it sounds: know which of your systems consume the output of another.

The Visibility Problem Gets Harder

Right now you have a quiet advantage you may not appreciate: you're still building AI systems consciously. Someone knows about each one. There's a project, a team, a deployment record. That is changing. As foundation models get easier to integrate, you will have systems running that no one in a governance role knows exist: a model one team fine-tuned for a narrow use case, a public model another team wired into a workflow, a cloud AI service a third team adopted that abstracts the model away entirely.

This is shadow AI, and it's the problem that keeps senior leaders awake: consequential systems in production that the governance function cannot see. Most governance programs are built on the assumption that they have visibility into what's being built. Tomorrow they won't. And the arithmetic is unforgiving. Governance covers the systems you know about; the ones you don't know about are, by construction, precisely the ones nobody reviewed. You can run flawless processes across your inventory and still be governing the safest half of your exposure.

The uncomfortable part is that shadow AI rarely looks like shadow IT. There's no rogue server and no one going around procurement. There's a SaaS tool whose vendor added an AI feature in a release note, and an analytics platform that shipped a forecasting module the finance team switched on. All of it arrived through channels you already approved, attached to products you already bought, and none of it will surface on a list you build by asking teams what AI they're running, because the people using it don't think of it as AI.

So the answer isn't a bigger approval gate; a gate only catches what walks up to it. It's a different architecture built around continuous discovery rather than up-front permission: instrumentation, monitoring, scanning for the signatures of AI (systems generating synthetic outputs, making pattern-based decisions, behaving like trained models). The organizations preparing now are already asking the operational questions. How would you find a deployed model you don't know about? What signal tells you a given system is using AI? Who goes looking when nobody has reported anything? Those are the capabilities the next few years will demand, and they're operational rather than philosophical.

Regulation Will Mature, and Create New Risk

Chapter 21 has already made the case that the rules will keep moving under you and that the durable response is a program you can reconfigure rather than rebuild. Take that as given. What matters here is what maturing regulation does to your *liability*.

Today, a good-faith governance program buys a lot of regulatory forgiveness. As the rules firm up, the standard shifts from *did you have a governance program?* to *did you have the right one?* And the second question is far harder to pass on sincerity alone. That is what makes governance drift expensive rather than merely untidy. A program that was genuinely state-of-the-art when you built it goes stale while you're still congratulating yourselves on having it, leaving you compliant with the rules you learned and non-compliant with the ones that replaced them, with no internal signal that anything changed. Forgiveness is a feature of an immature regime. It expires, and it expires without an announcement.

The Human Factor Matters More, Not Less

There's a comforting hope in governance: that once you have the right processes and frameworks, execution becomes routine, delegable, systematic. For AI, that hope is simply wrong, and it's wrong in the opposite direction from where intuition points. The more sophisticated AI governance becomes, the *more* it leans on human judgment. Is this model output surprising in a way that matters or an expected way? Is this emergent behavior a new risk or normal variation? When two governance requirements conflict, which wins? No process answers those. They require judgment, experience, and people who understand both the technology and the business well enough to see the connections.

So the organizations preparing well are investing in governance *people*: not compliance clerks, but people who can reason about risk, understand the technical substance of AI, translate fluently between engineering and the business, and are trusted by product teams as helpers rather than blockers. The instinct says mature governance should depend less on individuals. With AI, better governance requires better people. That's the counterintuitive part, and it's where a lot of programs will underinvest.

Trust Becomes the Battleground

Over the next few years a dynamic is likely to emerge: AI will start failing in ways that matter. Not catastrophically, not systems destroying companies, but visibly. Systems that discriminate. Systems that are confidently wrong. Systems that make a decision that, in hindsight, obviously should have been caught. Those failures will be reported, litigated, and felt in customer trust. And they'll hand a real opportunity to any enterprise that has genuine governance to point to.

An organization with real governance can say: we caught this; we understand our risks; we know where our systems can and can't be trusted; we're not perfect, but we're reliable. An organization without it is left saying: this was an unforeseen edge case, we have controls, we have policies (with nothing concrete to point to that caught the problem or would have). Trust becomes the battleground, and trust is won through demonstrated governance, not asserted governance. That distinction is the entire competitive story of the next wave.

The Work Doesn't End

If there's one thing worth carrying out of all of this, it's that AI governance is not a problem you solve. It's a problem you manage, continuously. The organizations winning are the ones that have accepted this: they aren't trying to build governance once and have it hold forever; they're building organizations that can think clearly about AI risk and adjust as the ground moves.

The challenges of the present (building frameworks, distributing expertise, scaling governance) are real, and they'll be met by enterprises that take the work seriously. The challenges ahead are different in kind: discovering systems you don't know exist, governing networks instead of standalone systems, staying ahead of regulation that will move under you, and building trust instead of governance theater. The organizations moving fastest already know this. They aren't waiting for perfect governance today. They're building governance that can absorb a landscape they can't yet see clearly. That is what it means to prepare for what comes next. And it is, fittingly, the least glamorous and most durable advantage you can build.

The Bottom Line

Compliance asks the wrong question. "What's the minimum we need to do?" optimizes for passing an audit, which is a low bar that governs nothing and locks in the adversarial, slow, incident-prone dynamic it was meant to prevent. Done as real infrastructure (distributed decision-making, predictive risk management, empowering frameworks, and learning that feeds back into architecture), governance stops being a tax on speed and becomes the reason you can move fast with confidence. The advantage isn't the policy binder; it's an organization that understands its own risks, and it compounds: the next wave of enterprise AI will be won on trust, and trust is earned through demonstrated governance, not asserted governance. Build now for the landscape you can't yet see (networks instead of standalone systems, discovery instead of gates, better people instead of more process), because governance is not solved once; it is managed forever.

Field Kit

1. The one-question diagnostic — bring it to your next leadership review. Ask: *Does governance slow us down or speed us up?* Then force the evidence. If every governance interaction adds time and the honest examples are all "we had to wait for approval," you have governance theater. If people can point to decisions governance made more confident, or remediation it prevented before deployment, you have infrastructure. Run it every quarter; the trend line matters more than any single answer.

2. Question set for the AI governance review. - What proportion of the AI systems in production can we actually name — and how would we know that number is wrong? - Where does one model feed another, and who owns the risk at the seam? - For which deployments could we prove — with concrete evidence, not policy language — that we caught a problem? - Which of our controls quietly assume the standard is still *did you have a program?* rather than *did you have the right one?* (Chapter 21 holds the forward-compatibility test itself.) - Is our governance team staffed with people product teams call for help, or people they route around?

3. The four-capability evidence test. For each capability below, name a concrete instance from the last quarter — a decision, system, or incident, with a date and a person attached — or score it absent. That it sounds like something you do is not evidence that you do it. *Distributed decision-making* — a team reasoned about AI risk and changed a design without a gate forcing them to. *Predictive risk management* — a signal got acted on before anything failed. *Empowering frameworks* — a deployment cleared on the first try because the criteria were knowable in advance. *Learning loop* — a failure changed how the next system got built, and you can point at the difference. The first absence is where the next investment goes.

4. Shadow-AI discovery starter. Answer three operational questions before your next planning cycle: (a) How would we find a deployed model no one told us about? (b) What signal would tell us a system is using AI — including one we bought, whose vendor added it in a release note? (c) Who owns continuous discovery — scanning, instrumentation, monitoring — as a standing function rather than a one-time audit?

5. The dependency trace. Take your three most consequential AI systems and draw what feeds them and what they feed — models, shared services, transformed data, decisions — two hops each direction. Then ask of every edge: if that component changed tomorrow, would anyone on this side notice, and whose risk register carries the seam? Edges nobody claims are where network failures start. If you can't produce the diagram at all, that is the finding.

6. The governance-hire test. Next time you open a governance role, read the posting against what the work demands: can this person reason about an unfamiliar trade-off, read the technical substance of a model, translate it into business terms, and be someone a product team calls early rather than late? If it reads like a req for a compliance clerk, you're staffing for the governance you have rather than the one you'll need. Then run the trust version on the team you have: when something went wrong last quarter, could you point to a practice that caught it — or only to policies that would have covered it?

The Field Kits

Every chapter ends with a Field Kit: a numbered set of workshop-ready tools you can put in front of a team without preparing anything first. There are 143 of them across the twenty-four chapters, and every one works as printed. The decision records, the RACI grid, the readiness audits, the board risk dashboard, the unit-economics scorecard and the risk-appetite templates are all here, in the book, in the chapter whose argument they belong to.

That placement is deliberate. A tool separated from the reasoning behind it becomes a form somebody fills in, and a filled-in form is not governance. Read the chapter, then run its Field Kit in the room where the decision actually gets made.

How to Bring This Into Your Organization

Reading a governance book alone rarely changes a governance system. The change happens when the people who hold the decision rights sit in the same room and work through the material together, which is what the artifacts in this book were designed for.

Three formats work, in ascending order of depth:

The ninety-minute executive briefing. For a board, an executive committee, or a leadership team that needs to understand the accountability gap and leave with one decision made. Usually the decision is: who owns AI deployment approval, and what are they accepting when they sign.

The full-day governance design workshop. For a cross-functional group (risk, legal, data, technology, the business) that needs to actually build the operating model. The group completes the three-level decision sort, the operational-readiness audit, and the board risk dashboard for their own real systems. They leave with a working draft, not notes.

The multi-week enablement series. One session per part of this book, for organizations building governance capability rather than buying it. This is the format that changes how an organization thinks, because the material lands against real systems over time.

You do not need anyone's help to run any of these. The artifacts are designed to be facilitated by someone inside the organization who knows the politics, which is usually the person best placed to do it anyway.

About the Author

Bandar Alsharfi (most people call him Ben) works on scaling artificial intelligence in complex and highly regulated environments, and is drawn to the parts of it that do not demo well: governance, compliance, cost, vendor risk, business value, and the question of who signs off when something goes wrong.

He has spent his career in enterprise technology, and serves as a non-executive director and board member with oversight of a portfolio of companies internationally. That puts him on both sides of the question this book asks: the person recommending a system, and the person on the board who has to answer for it. He holds an Executive MBA and a master's degree in AI engineering.

He writes at **codedebate.com**, where this book began as *In Practice: AI in the Enterprise*, a ninety-part series published every working day. It set out to document something he kept encountering and could not find written down honestly anywhere: that enterprise AI rarely fails on the technology, and almost always fails on the question of who decided, on what basis, and who is watching now.

He is most useful to organizations at the point where technical reality meets executive accountability — translating between the teams who build these systems and the people who must answer for them.

The views expressed in this book are the author's own and do not represent those of any organization with which he is affiliated.

The Fifteen Pillars: Quick Reference

The governance pillars this book returns to, grouped as they function. Each pillar's question is its contribution to its cluster.

CHOICE ARCHITECTURE	THE SET OF MOVES ACTUALLY AVAILABLE TO YOU
Decision rights	How fast can we move?
Cost control	What are we allowed to decide?
Vendor strategy	What options are even on the menu?

RISK ENVELOPE	THE BOUNDARY OF WHAT YOU CAN SAFELY HOLD
Model risk	What could go wrong?
Data quality	How much risk are we actually carrying?
Architecture	Can we change our mind later?

ACCOUNTABILITY SURFACE	WHO IS RESPONSIBLE, AND WHETHER THAT MAP IS REAL
Accountability	Who answers when it breaks?
Board oversight	Has an appetite been set?
Compliance	Would a regulator accept this?

INVESTMENT DISCIPLINE	WHETHER THE MONEY TURNS INTO ANYTHING
Capability	How much do we do ourselves?
Adoption	Did anything actually change?
Measurement	Can we tell?

DEPLOYMENT RELIABILITY	WHAT HAPPENS WHEN SOMETHING MOVES
Change management	How do updates ship?
Monitoring	Will we find out?
Operational risk	What happens next?

A sixth cluster, **strategic flexibility**, adds no new pillars: architecture, vendor strategy, and model risk again. Each anchors two clusters, so tuning one moves the others.

Further Reading

Rather than a bibliography, these are the primary sources worth reading directly. Frameworks and regulations evolve; go to the source rather than to commentary about it.

Regulation and standards

- The EU Artificial Intelligence Act: the first comprehensive statutory regime, and the reference point most multinational programs are calibrating against.
- Regulation (EU) 2026/1744, the Digital Omnibus on AI: the first set of amendments to the AI Act, and a useful record of what happens to a compliance timetable under implementation pressure.
- NIST AI Risk Management Framework (AI RMF): voluntary, practical, and the closest thing to a common vocabulary for AI risk in the United States.
- ISO/IEC 42001: the AI management system standard; useful if your organization already runs certified management systems.
- ISO/IEC 23894: AI risk management guidance.
- Sector-specific supervisory guidance in your own industry. For regulated firms this usually matters more than any general AI regulation, and it is frequently overlooked.

Adjacent disciplines worth borrowing from

- Model risk management practice in financial services, and SR 11-7 in particular: the 2011 Federal Reserve and OCC supervisory guidance that most enterprise model governance is descended from. Decades ahead on validation discipline, and discussed in this book as both a resource and a trap.
- Safety and incident-investigation literature from aviation and healthcare: the source of the blameless-review approach in Part II.
- Internal audit and assurance standards from the professional bodies: increasingly the function being handed AI assurance without a playbook.

A note on citation. This book deliberately makes no predictions about where regulation lands. Every case in it is a composite, assembled from patterns observed across many organizations rather than drawn from any one of them; any resemblance to a specific organization's circumstances is a reflection of how common these patterns are, not a description of that organization. Regulatory dates cited in this book are accurate as at July 2026 and are used to illustrate volatility, not as a compliance reference. Verify current obligations against primary sources.

Index

Page references are to the printed page. Entries are grouped so the book's structure stays visible: the fifteen pillars, the six clusters they form, the recurring concepts, and every Field Kit tool by its own name.

THE FIFTEEN PILLARS

THE SIX CLUSTERS

CONCEPTS AND TERMS

FIELD KIT TOOLS

www.ingramcontent.com/pod-product-compliance
Lightning Source LLC
LaVergne TN
LVHW052037200926
848481LV00005B/1710